HANDPRINTS ON HUBBLE

LEMELSON CENTER STUDIES IN INVENTION AND INNOVATION
Joyce Bedi, Arthur Daemmrich, and Arthur P. Molella, general editors

Arthur P. Molella and Joyce Bedi, editors, *Inventing for the Environment*

Paul E. Ceruzzi, *Internet Alley: High Technology in Tysons Corner, 1945–2005*

Robert H. Kargon and Arthur P. Molella, *Invented Edens: Techno-Cities of the Twentieth Century*

Kurt Beyer, *Grace Hopper and the Invention of the Information Age*

Michael Brian Schiffer, *Power Struggles: Scientific Authority and the Creation of Practical Electricity Before Edison*

Regina Lee Blaszczyk, *The Color Revolution*

Sarah Kate Gillespie, *The Early American Daguerreotype: Cross-Currents in Art and Technology*

Matthew Wisnioski, Eric S. Hintz, and Marie Stettler Kleine, editors, *Does America Need More Innovation?*

Kathryn D. Sullivan, *Handprints on Hubble: An Astronaut's Story of Invention*

HANDPRINTS ON HUBBLE

AN ASTRONAUT'S STORY OF INVENTION

KATHRYN D. SULLIVAN

THE MIT PRESS
CAMBRIDGE, MASSACHUSETTS
LONDON, ENGLAND

First MIT Press paperback edition, 2020
© 2019 Smithsonian Institution

All rights reserved. No part of this book may be reproduced in any form by any electronic or mechanical means (including photocopying, recording, or information storage and retrieval) without permission in writing from the publisher.

This book was set in Arnhem Pro by The MIT Press. Printed and bound in the United States of America.

Library of Congress Cataloging-in-Publication Data

Names: Sullivan, Kathy, 1951– author.
Title: Handprints on Hubble : an astronaut's story of invention / Kathryn D. Sullivan.
Description: Cambridge, MA : The MIT Press, c2019. | Series: Lemelson Center studies in invention and innovation series | Includes bibliographical references and index.
Identifiers: LCCN 2019005641 | ISBN 9780262043182 (hardcover : alk. paper) 9780262539647 (paperback)
Subjects: LCSH: Sullivan, Kathy, 1951– | Women astronauts—United States—Biography. | Women scientists—United States—Biography. | Hubble Space Telescope (Spacecraft)
Classification: LCC TL789.85.S85 S85 2019 | DDC 522/.2919—dc23 LC record available at https://lccn.loc.gov/2019005641

10 9 8 7 6 5 4 3

CONTENTS

SERIES FOREWORD

Space: the final frontier. For thirty years, our window into that vast unknown has been the Hubble Space Telescope. The images it has sent back to Earth are iconic. They have inspired our imaginations and deepened our awe of what lies beyond our home planet.

Hubble's reputation wasn't always so stellar. Its first images were horribly blurry. The telescope was mocked and berated in the media. Congress held hearings into what went wrong with the $2 billion project. But the audacity—and success—of the mission that fitted Hubble with a "contact lens," as the contemporary press likened the repair, became one of NASA's most glorious and well-known victories.

There is much more to Hubble, however. In this book, Kathryn Sullivan—former astronaut, the first American woman to walk in space, and a member of the space shuttle crew that launched the telescope—tells the little-known story that Hubble, from its first moments on the drawing board, was designed to be maintained, repaired, and upgraded. *Handprints on Hubble* chronicles her first-hand experiences as part of the team that invented and perfected the tools that made on-orbit maintenance possible.

As this book goes to print, a growing number of influential scholars are focusing attention on the importance of maintenance in a technological society, and arguing against what they see as a one-sided emphasis on innovation. Without doubt, innovation has become a universal watchword and paramount objective of the twenty-first century. Yet, maintenance and innovation need not be at odds. Instead, as the tools that were invented to maintain the Hubble Space Telescope illustrate, effective maintenance requires new inventions and can spur more widely adopted innovations.

Invention and innovation have long been recognized as transformational forces in American history, not only in technological realms but also in politics, society, and culture. Since 1995, the Smithsonian's Lemelson Center has been investigating the history of invention and innovation from an interdisciplinary perspective. Books in the Lemelson Center Studies in Invention and Innovation extend this work to enhance public understanding of humanity's inventive impulse. Authors in the series raise new questions about the work of inventors and the technologies they create, while stimulating cross-disciplinary dialogue. By opening channels of communication between the various disciplines and sectors of society concerned with technological innovation, the Lemelson Center Studies advance scholarship in the history of technology, engineering, science, architecture, the arts, and related fields and disseminate it to a general interest audience.

JOYCE BEDI, ARTHUR DAEMMRICH, AND ARTHUR MOLELLA
Series Editors, Lemelson Center Studies in Invention and Innovation

PROLOGUE

April 24, 1990, found us right back where we had been fourteen days earlier: suited up, strapped in, and ready to go, with the countdown clock stopped at T-31 seconds. Again. This time the launch control center computers had halted the countdown because of an indication that a valve on one of the pipes used to fill the external fuel tanks had failed to close. If the indicator was correct, then only one valve was left to prevent the fuel in the tank from leaking overboard instead of feeding into the space shuttle's three main engines. If that happened, we could end up too low to deploy the Hubble Space Telescope, at an abort landing site on the other side of the Atlantic, or splashed into the ocean. The launch would be scrubbed rather than accept that risk. If the indicator was wrong, however (think of the flaky tire pressure sensor on your car), then the engine system was fine and there was no reason to scrub. Which was it: Serious problem, or faulty indicator? Go for launch, or scrub?

This high-stakes call fell to the launch team controller responsible for the shuttle's main propulsion system, someone I still know only by the call sign "MPS." Time was not on his side. The shuttle's auxiliary power units set a strict limit on how long we could hold at this point: only twelve minutes more. In the cockpit, we listened intently as the launch team worked out the problem. "MPS, what's your status?" the launch director asked. The propulsion engineer talked calmly through the data on his display. The temperature and pressure readings in the surrounding lines were not consistent with an open valve; fundamental physics said it was closed. He proposed to send a manual command for the valve to close, hoping this would make the indicator read correctly. It worked, but the control center computers still had a lock on the countdown clock. "MPS, what is your call?" the launch director pressed. "I am prepared to manually override the software and proceed with the count," he replied. With a crisp and rapid cadence the best soldier would envy, the launch director gave him the GO to do that and told the other launch controllers to get ready to continue the countdown. Then he advised the NASA Technical Director (call sign "NTD"), who alone could restart the countdown clock in a situation like this, that the launch team was again GO. The call we had been hoping for came a split second after: "All controllers, this is NTD. The countdown clock will resume on my mark. Three, two, one, MARK." The entire episode had taken less than three minutes. Aboard the shuttle, we marveled aloud at the technical acumen and cool professionalism the MPS engineer had shown as he saved us from a scrub by diagnosing the problem so swiftly, and beamed with delight as the countdown clock again began to tick down toward zero.

Thirty-one seconds later, *Discovery* roared off the launch pad. Sitting on the lower deck, with nothing but a wall of storage lockers to look at, I closed my eyes and took in the sounds and sensations of a space shuttle launch. The solid rocket motors, which are essentially gigantic firecrackers, made the first two minutes and fifteen seconds turbulent and loud. I felt like I was in an earthquake and a fighter jet at the same time. The vibrations were almost bone-rattling; the thrust pushing through my back strong and constant. I felt the thrust tailing off just before Charlie announced that the chamber pressure in the solid rockets was decreasing, then heard the *thump* that confirmed they had been jettisoned. The ride was much quieter now, and as smooth as an electric train. The push against my back continued, as thrust from the main engines accelerated us toward orbital velocity. Six minutes later, they cut off as planned. The lightness in my arms and legs and the checklists floating at the ends of their tethers announced that we were in orbit. Although nearly six years had passed since my first spaceflight, I felt instantly at home.

ACKNOWLEDGMENTS

"The only thing we do truly alone is have the start of a good idea" is one of my favorite adages, a reminder of how much we depend on other people in life. Certainly, this book would not have been possible without the help and support of a great many people from the Smithsonian Institution, NASA, and that wonderful extended family known as Team Hubble. I am more grateful for their help and support than the few words on these pages can express.

The launch pad for this work was a Lindbergh Fellowship in Aerospace History at the National Air and Space Museum. Special thanks to my longtime friend Valerie Neal for bringing that opportunity to my attention and supporting my application. Joyce Bedi, at the Smithsonian's Lemelson Center for the Study of Invention and Innovation, became an early champion and connected me to Katie Helke at the MIT Press. Joyce and Katie's editorial guidance and encouragement throughout this journey have been invaluable.

I could not have reconstructed the early history of the Hubble Space Telescope and the development of the maintenance concept without the help of expert historians and archivists. My colleagues in the Space History Department at the National Air and Space Museum, notably Valerie Neal, David DeVorkin, and Margaret Weitekamp, helped ground me in basic historical methods and connected me to useful sources. Liz Borja and Brian Haskins schooled me on archival research practices and cheerfully and efficiently tracked down items in the museum's Space Telescope History Project collection, often delivering gems I had not known to seek out. I also thank Colin Fries and Elizabeth Suckow at NASA headquarters and Brian Odom at the Marshall Space Flight Center for their assistance with those historical collections.

The great gift of this project was the joy of reconnecting with so many of my former Hubble colleagues. All were eager to reminisce and excited that this overlooked part of the Hubble story would finally be told. They were also unfailingly patient with the countless emails I sent as I labored to reconstruct the historical timeline and understand events that had happened out of my sight. I am deeply grateful for their friendship and assistance. The responsibility for any and all errors in the final product is, of course, mine alone.

Special thanks go to Ron Sheffield, Bob O'Dell, Peter Leung, and Brian Woodworth, who welcomed me into their homes, opened their personal collections, and patiently fielded innumerable questions. Lockheed engineers Frank Costa, Tom Dougherty, Lee Dove, Julie Sattler, Tom Styczynski, and Dom Tenerelli helped me reconstruct the company's early maintainability design efforts and the timeline of hardware development. Jean Olivier and John Reaves were equally generous in answering my questions about the early history at the Marshall Space Flight Center. At the Johnson Space Center, Robert Trevino and Michael Withey gave me valuable new insights about the inventive process of EVA equipment development, and Jim Skipper filled in long-forgotten details about our STS-31 vacuum chamber tests.

Many colleagues helped fill gaps in my memory about the Hubble deployment and servicing missions, including my STS-31 crewmates Loren Shriver, Charlie Bolden, Steve Hawley, and Bruce McCandless; flight director Bill Reeves; and EVA specialists Sue Rainwater, Jim Thornton, and Robert Trevino. My TFNG classmate Jeff Hoffman, a spacewalker on the first Hubble servicing mission, along with Frank Cepollina and Pat Crouse at the Goddard Space

Flight Center, provided valuable recollections, documents, and photographs about the tools, operations, and lessons learned from all five servicing missions. Carol Christian guided me to key illustrations, facts, and figures within the Space Telescope Science Institute's vast holdings. Kate Carroll Bulson transcribed the interviews I recorded with my colleagues and the STS-26 wakeup music lyrics that Mike Cahill very graciously allowed me to reproduce. I thank them all for their efforts on my behalf.

Constructive feedback is indispensable on the long and arduous journey from concept outline to final text. I owe a special debt of gratitude to David DeVorkin, Valerie Neal, Andy Plattner, Renee Stone, and Kitty Williamson, who generously agreed to read very early drafts. Critical reviews by Michael Lewis, Susanne Jaffe, Charlie Justiz, and Bill Readdy helped turn those early versions into a final manuscript.

The final salute must go to my parents, Don and Barbara Sullivan, who stoked my curiosity, taught me to fly, and encouraged me to reach for the stars.

ACRONYMS

EVA	Extravehicular Activity, or spacewalk
M&R	Maintenance and repair/refurbishment
NASA	National Aeronautics and Space Administration
NBS	Neutral Buoyancy Simulator (the water tank at Marshall Space Flight Center)
ORU	On-orbit Replaceable Unit
PFR	Portable Foot Restraint
STHP	Space Telescope History Project
STS	Space Transportation System
TFNG	Thirty-Five New Guys (the nickname for the astronaut class of 1978)
VATA	Vehicle Assembly and Test Area
WETF	Weightless Environment Test Facility (the water tank at the Johnson Space Center)

1

INTRODUCTION

"Tell us about yourself. Start with high school."

So began the most momentous interview of my life. I looked at the fourteen experienced space professionals sitting around the conference table in front of me as I pulled my thoughts together. Four of them were astronauts whose adventures I had followed closely over the years. I could never have imagined at the time that I would someday be at the Johnson Space Center in Houston, Texas, talking with them about the possibility of joining their ranks. The next ninety minutes would turn the twenty-six-year-old fledgling oceanographer I was on that November day in 1977 into an astronaut and put me on a collision course with another notable spacefarer, the Hubble Space Telescope.

It is our most celebrated eye in the sky. Type the name "Hubble Space Telescope" into a search engine and you will get over 14 million hits. Hubble and its images of stars and galaxies have become part of our pop culture, appearing on everything from calendars, tote bags, and U-Haul trucks to human limbs (figure 1.1).

FIGURE 1.1

A few examples of how the Hubble Space Telescope appears in everyday life. Source: NASA.

Hubble has allowed us to peer deeper than ever before into the cosmos and has revolutionized our understanding of the universe around us. It has looked into the stellar nurseries where stars are born; revealed thousands of galaxies in what seemed to be empty patches of sky; transformed our understanding of black holes from super-massive objects in galaxies to star-sized objects in stellar clusters; found dwarf planets and protoplanetary disks—the clouds of matter from which planets form—around other stars; measured precisely how fast the universe is expanding; and discovered water plumes on Jupiter's moon Europa. We felt like participants in these discoveries—almost as if we were in that distant star cluster—thanks to the stunning images that paraded across our televisions and computer screens.

The Hubble Space Telescope has done something that no other scientific satellite has ever done: it has evolved while in orbit, getting better as it aged. The telescope orbiting the Earth today is a vastly more powerful and more efficient scientific instrument than the one my crewmates and I put into orbit back in 1990. This improvement would not have been possible without the telescope's inherently maintainable design and the maintenance capacity of the space shuttle and her astronaut crews. Today's Hubble produces 20 percent more power with solar arrays that are one-third smaller than the original ones. Its reliability, data storage, and data transmission rates all rose as space shuttle crews replaced the original vintage-1970s electronics with state-of-the-art solid-state components. Maintenance paid its most valuable dividend in greater scientific capability: Hubble's cameras are hundreds of times better today than they were at the outset, and its spectrographs cover more wavelengths with much greater sensitivity. All these advances allowed Hubble to measure the expansion rate of the universe, known as the Hubble Constant, almost five times more precisely than the preflight design goal (to less than 2 percent versus the original goal of 10 percent) and to become recognized as the most productive observatory ever built.[1]

✷

You could say we grew up together, Hubble and I. Our life stories began at about the same time. I was born to Donald and Barbara Sullivan in Paterson, New Jersey, on October 3, 1951. Most people trace Hubble's start to a report written in 1946 by Princeton astronomer Lyman Spitzer, titled "Astronomical Advantages of an Extra-terrestrial Observatory."[2] As I grew through my childhood and teen years, Hubble languished in a state of arrested development, while waiting for the technologies of the new space age to mature enough to bring Spitzer's vision into the realm of the possible. We both then spent several years spent figuring out who or what we would turn out to be when fully grown. For me, this stage was called high school and college, and the outcome was a budding oceanographer with a hunger for travel and adventure. For Hubble, it went by the NASA labels "Large Space Telescope Pre-Phase A" and "Phase A," and the outcome was a conceptual design for a spacecraft that could fulfill Spitzer's vision.[3] I went on to graduate school in 1973, where I acquired the deeper knowledge and broader experience that would lead to a radical academic course change and eventually my selection as an astronaut. That same year, Hubble went on to Phase B design, in which engineers developed the technical detail and analytical rigor needed to persuade both NASA and Congress to approve construction of a real spacecraft. By delightful coincidence, the two of us began the final stretch of our journeys to space at almost the same time, with my astronaut selection and Hubble's congressional approval both occurring in 1978.

We collided in early 1985. My boss George Abbey, then head of Flight Crew Operations at the Johnson Space Center, summoned me to his office to tell me that fellow astronaut Bruce McCandless and I would be the spacewalkers on the shuttle mission slated to deploy the Hubble Space Telescope in October 1986. On the surface, this looked like a straightforward satellite deployment mission that would use the shuttle's manipulator arm. That was not the type of mission that would usually call for assigning spacewalking experts nearly two years in advance, but there was much more to this story.

Hubble needed to operate and stay abreast of technology for fifteen years to fulfill the ambitious science agenda that justified its high cost and won it political support. As far back as 1965—before Russian cosmonaut Alexey Leonov had floated out of his capsule on the world's very first spacewalk—the strategy for achieving this was to have astronauts maintain and upgrade the telescope in orbit. Our early assignment reflected the fact that the tools, equipment, and operating procedures needed to do this were nowhere near ready. The choreography for some tasks had been roughed out in a few zero-gravity water-tank simulations, and two earlier shuttle flights had tested some potential tool designs, but little to no work had been done on many other tools and tasks. Our job from that meeting until launch day was to drive this work forward and ensure that we had reliable spacewalking equipment and procedures, not only for our deployment mission but also for the much longer list of maintenance tasks that would keep Hubble alive and at the cutting edge scientifically. This would take lots of simulated spacewalks (which NASA calls EVA, for "Extra-Vehicular Activity"[4]) in the neutral buoyancy simulator (i.e., immense water tank) at the Marshall Space Flight Center in Alabama (henceforth Marshall), as well as many trips to the plant in California where Hubble was being built, to prove that the equipment we were developing worked on the real telescope and not just the training mockups. During our flight, the deployment mission, Bruce and I would be spring-loaded to jump into our spacesuits and head outside to fix any one of half a dozen problems that could be fatal to the telescope.

Even without an assured EVA, I considered it a dream assignment.

The adventure I lived from that day in my boss's office in 1985 until the day we placed Hubble in orbit in 1990 is the raison d'être for this book. My motive in writing it is twofold, one part historical and one human. On the historical side, I aim to fill in an overlooked chapter in Hubble history. It was during those five years that the general notion of tending a telescope in orbit became a practical reality. On the human front, my goal is to shine a light on some of the unsung engineers who made this happen, inventing, producing, and testing the concrete things—tools, support equipment, and operating procedures—that it takes to maintain such a complex machine in space. These people, along with the engineers who designed maintainability into the telescope from the start, are truly Hubble's first maintainers. The sterling reputation the telescope enjoys today rests on the maintenance capability they established and passed down to the five space shuttle servicing crews who visited Hubble between 1993 and 2009. Bruce and I worked hand in glove with this unsung band of engineers, known as "the M&R team" (for "maintenance and repair"), as we prepared for our mission to deploy Hubble and laid the foundation for the maintenance missions crews that would follow us.

My role in the pages that follow is a mixture of participant, narrator, observer, and amateur historian, sharing both what I experienced as a young astronaut seeing the world around me through a very tiny peephole and what I have learned since then about the tapestry of events in which I was caught up. The narrative is organized loosely around my life story, looping back in time to fill in historical context where needed and cutting between my own experience and noteworthy contemporary events. Finally, a note on voice. Many have remarked that I almost always say "we" rather than "I" when I talk about my work as an astronaut. This is neither false modesty nor some Pollyanna-ish NASA brainwashing about teamwork. It simply reflects how I saw things and how I went about my work. I played on some great teams at NASA, but none as fine as the Hubble M&R team. True, I had a prominent position and played it well, but my success still relied on much from other people. My colleagues on the M&R team have all been delighted that the story of our work together will finally be told, but also reluctant to be called out above the others. Even my crewmate Bruce McCandless shied away from claiming sole credit on initiatives he spearheaded and devices he patented. There are many styles of teamwork. Ours was driving and demanding on the technical side, but remarkably constructive and mutually supportive on the human side. What follows is every bit as much the team's story as my own.

2

SPUTNIK BABY

I felt like I was joining the world's greatest adventure movie—already in progress—when I arrived in Houston as an eager young astronaut in 1978. Growing up during the earlier segments of this movie, I had been keenly aware that I was watching a group of people do something amazingly complex and daring that nobody had ever done before. The aura of supreme expertise and unflappable confidence projected by the dashing men involved (always men back then) was alluring, and the tales and photos the early astronauts brought back entranced me. I pored over every magazine story about these people and their exploits and glued myself to our television whenever a mission event was broadcast. As I waited backstage to be formally introduced to the media gathered in the Johnson Space Center auditorium, I could hardly believe I was shifting from spectator to participant in the grand adventure of spaceflight.

My path to this moment had been rather unusual. Maps were my earliest childhood fascination, not airplanes and rockets. Maps revealed fascinating new things about places I knew and fired my imagination about foreign places and the people who lived there. The map that came with each issue of *National Geographic* disappeared into my bedroom the moment the magazine arrived at our house, added to the treasure trove I hid under my bed. By the time I was seven, I knew how to read map coordinates and had drawn my first map. I shocked my parents that same year by plotting, accurately and without any help, the driving route we should take to a faraway vacation destination in the Sierra Nevada Mountains.

So, although the early spaceflights entranced me and I devoured everything written about the missions and the astronauts, the thought of becoming an astronaut never entered my mind (except for wondering what it would feel like to see the Earth from space with my own eyes). Instead, these bold exploits fueled a deep but vague longing for a life that was filled with similar grand adventures.

When I discovered at about age ten that I had a flair for foreign languages, I figured this would be my ticket to that kind of life. My simple theory was that I would learn a lot of languages and somehow parlay my talent into a job that involved exploring the world and living in exotic lands. James Michener's *Caravans* had shown me there was at least one such job: Foreign Service Officer. Surely there were others, I thought, though I had no idea what they might be.

*

My world turned upside down when my maternal grandmother died in March 1963, when I was twelve. I was confused and saddened by her passing, but my mother, I would soon discover, was broken.

Pauline Dwyer Kelly was the only one of my four grandparents I ever knew. Both of my grandfathers had died before I was born, and my other grandmother followed them before I turned two. Nana abandoned the cold of New York City to join us in southern California a few years after we moved there from New Jersey. Her small apartment became a common stop for me and my friends on our neighborhood expeditions—an oasis where we could cool off and refresh with cookies, milk, and her wonderful hugs. My older brother Grant and I loved to have her babysit us, unless she brought her Golden Age Club friends along with her. We wanted to play with Nana ourselves, not listen to her friends' laughter and their shouts of "Gin!" or "Yahtzee!" She ended the Golden Age parties as soon as we told her that.

My mother and I had stopped by her apartment one December day in 1962 to drop off some groceries. Out of the blue, Mom bolted from the kitchen and into Nana's bedroom. The next thing I knew we were at the hospital. The doctor said it was a stomach ulcer, but only because he did not want to burden us with a cancer diagnosis right before what was sure to be her last Christmas.

The strong and nurturing mother I had known since infancy disappeared soon after Nana's funeral. Overwhelmed by an anguish that was beyond my comprehension, she descended into depression and drink. My father struggled to understand what had befallen his wife and how to help her, while juggling the needs of his two bewildered children and continuing his work as an aerospace engineer. My brother and I tried to protect her when he was at work, disabling the car so she could not drive and devising ploys to hide her drinking from visitors.

My biggest fear was that my father would become so worried about her condition and its effect on us kids that he would send her away. "Away" to what or where, I did not know, but her fear of this scenario and the prospect it raised of shock treatment was palpable to me. Although I had essentially lost my mother already, I was sure she would be gone forever if she was sent away. My reactions to the loss and fear of this prospect followed a classic pattern: I appointed myself housekeeper and tried to be the always-strong-and-calm intermediary, ever-watchful and ready to stabilize tense situations or assuage everybody else's pain and anger. After-school sports and Girl Scouts became my escapes from the tumult and strain at home.

Though she never was sent away, she also never came fully back to us. The passage of time and many hours of therapy eased her depression and brought an end to her destructive drinking, but never completely restored the mother I had known as a child. I would catch glimmers of her former self when I watched her play with younger neighborhood kids she invited over for a swim in our pool. I would have given anything for a time machine that could take me back to when I was the child sharing such simple, joyful moments with her. There was no time machine, of course; our world was forever altered, and me along with it. Over time I would move past the anger I felt at first and come to appreciate the inner strength and independence these turbulent years forced me to develop.

✷

By age fourteen, I was taking both French and German in school and had set my first concrete life goal: getting into a college with a junior-year abroad program, so that I could live in a foreign country. All seemed to be going my way when I was admitted to the University of California at Santa Cruz (UCSC). Known mainly for its beautiful campus set on the hills overlooking Monterey Bay and, back then, its hippie ways, the main draw for me was UCSC's outstanding Russian language program.

As I walked into the first meeting with my academic advisor, Professor of French Literature John Hummel, the most pressing questions on my mind were how quickly I could start Russian classes, which courses would prepare me best for a career in the Foreign Service (still the only specific job I knew of), and whether there was some way I could take more than the three classes per term that were then standard at Santa Cruz. I had spent hours poring over the course catalog during the summer and knew I couldn't possibly fit everything I wanted to take into four years at the measly rate of three classes per term. John smiled kindly before informing me that there was a much more immediate issue to decide: which three science classes I would take to fulfill my freshman year general education requirement. Having helped many a young language major clear this hurdle, he suggested three that he knew to be interesting, well taught, and not too hard. I argued fiercely against what I considered a needless delay on my path to Russian fluency and my dream life of exploration and adventure.

Luckily for me, I failed to win the argument. The mandatory science courses foisted on me that day introduced me to the earth sciences—geology and oceanography, specifically—and to young faculty scientists who were living exactly the kind of inquisitive and adventurous life I had dreamt of for so many years. More excited than ever about my studies, I changed majors at the end of freshman year and set out to become an oceanographer.

This course change eventually took me to Bergen, Norway, for my long-awaited junior year abroad, and then to Nova Scotia, for graduate work at Dalhousie University and Canada's premier oceanographic center, the Bedford Institute of Oceanography.

I loved everything about oceanography, but especially going to sea. A successful research expedition seemed like a symphony to me. Both are unique compositions that weave together many strands—the varied orchestral voices in one case, and the science, the ship, and the sea in the other. A lot of hard work, deep thought, careful analysis, and fierce debate go into planning the cruise beforehand—writing the score, as it were. Finishing a cruise plan always left me feeling elated about our creation, relieved that the ordeal was over, and exhausted, just as I imagine a composer must feel when a musical score is done. Then you must bring it all to life in the performance. The conductor raises her baton; the ship slips her mooring lines and heads out to sea. Once you're underway, only two things are certain: there is no chance of success unless everyone plays their part well, and there is no way things will unfold exactly as planned. Improvisation plays a crucial role in determining the outcome in both cases. And, at sea as in music, great improv requires mastery of the art and its tools. I found it all exhilarating and leapt at every chance to go out to sea.

✷

I could see the light at the end of the PhD tunnel when I went home to California for the Christmas holidays in 1976. A long stretch of number crunching, map drawing, and dissertation writing was still ahead, but I was beginning to think about what I wanted do after I finished my degree. The highest item on my wish list was to become a scientist aboard the *Alvin* submersible, operated by the Woods Hole Oceanographic Institution (figure 2.1), so that I could see the volcanic landscapes of the seafloor in person. Another intriguing prospect was to shift gears and pursue the emerging new field of satellite oceanography, which was already changing the way we understood the oceans. Either of those would give me another intriguing perspective on how the Earth works. I had investigated a number of junior faculty openings and postdoctoral fellowships at that point, but not yet found an opportunity that would open one of these doors.

FIGURE 2.1

The deep submergence vessel *Alvin* descending into the ocean. Photo by Chris Linder. Source: Woods Hole Oceanographic Institution.

My brother Grant tossed out a third prospect while we were talking on Christmas afternoon. NASA was developing a new and very different kind of spaceship, a reusable space shuttle, and they were looking for scientists and engineers to conduct the wide array of missions envisioned for the craft. Grant was the real flying nut in our family. One of my earliest memories is of him at about age four peering through a chain link fence at the small planes parked at an airport near our home in New Jersey. He had earned his pilot's license shortly after finishing high school and completed an undergraduate degree in aerospace engineering before landing his dream job as a corporate jet pilot. Grant had already sent two applications to NASA, one to be a shuttle pilot and the other to be a mission specialist, as NASA called the scientists and engineers who would make up the shuttle's crews. They were working hard to recruit women and minorities, he noted, adding that he thought I would have a very good chance of being selected. After all, he said, "How many twenty-six-year-old female PhDs can there be?"[1]

My first reaction was to dismiss the idea out of hand. It is hard enough to study the ocean floor through thousands of feet of seawater. Adding another two hundred miles of separation to the challenge was absurd. After returning to Nova Scotia, I saw a NASA recruiting advertisement in a science journal that made me think again about Grant's silly suggestion. It dawned on me as I read the advertisement that NASA was building a research ship. A very different and wildly faster vessel than any I had ever worked on, to be sure, but a research ship nonetheless. I was really good at running expedition operations at sea, so why not use my skills in space?[2] I would probably have to give up my long-standing dream of diving down to the seafloor in a submersible if I was selected, but getting to see the Earth from space with my own eyes instead seemed a more than fair exchange. I sent away for the application materials, filled out the incredibly long civil service forms, and dropped the packet into a Canada Post box.

NASA fell out of my mind the instant my application hit the bottom of the post box. I dove back into the all-consuming world of my PhD work. There was a final research cruise to plan for the upcoming North Atlantic field season, scads of data to analyze, and many maps to make and puzzle over. Only then would I be able to piece together the geological history of the Newfoundland Basin, my small corner of the Northwest Atlantic Ocean, and only then could I know the punch line of my dissertation.

Astronaut wasn't the only intriguing job prospect I lost sight of during that year's all-consuming work. Lying unanswered amid the mountain of paper that passed for my desk was a letter offering me the oceanographic opportunity I had been aiming at for years: exploring the deep seafloor in person aboard the *Alvin*. The sender was Dr. William B. Ryan of Columbia University's Lamont-Doherty

Geological Observatory,[3] one of the luminaries in my field of marine geology. Bill called sometime in October to ask if I intended to accept his postdoctoral fellowship position or not. I told him I was thrilled by the offer but needed to hear back on one other prospect before I could make a final decision.

Our conversation took an unexpected turn when I told him what that prospect was. A decade earlier, Bill had been a finalist in NASA's selection of a second batch of scientist astronauts. Not making the final cut had turned out to be a blessing in disguise for him. The class he missed out on joining arrived at the agency just as the Apollo program began to wind down. Their nickname became the "Excess Eleven," and those who had stayed at NASA were still waiting for their first spaceflight. His experience gave him a savvy and sympathetic perspective on my situation, which he shared with me generously. I agreed to call him back after finding out where I stood in the selection process and when NASA expected to make final decisions.

✷

"Haven't we told you no already?" asked the man who answered the phone in the Johnson Space Center's astronaut selection office. He sounded surprised when I said that they had not told me anything at all, and put me on hold to go find my file. "You're scheduled for an interview next month," he said, when he came back on the line. "You'll get a telegram sometime next week with all the information." I pressed him for some insight about the interview's significance and my chances of selection, but he gave only general answers. He was more forthcoming about schedule: NASA hoped to announce the new class by year's end.

The invitation to go to Houston for an interview meant that my odds of becoming an astronaut had improved from one in many thousands to one in a few hundred, explained Dr. Ryan on our next call. If my weeklong stay in Houston mirrored his earlier experience, it would include a barrage of medical and psychological examinations and a lengthy face-to-face interview with a panel of veteran astronauts and senior Johnson Space Center officials. He was clearly excited for me and agreed to postpone the fellowship decision until the NASA process had run its course. "We both know the odds are that you are coming to Lamont," he said, "but this is one you don't walk away from voluntarily. You make them tell you 'no.'"

✷

"So, what exactly does this mean?" my mother asked when I shared the news of my NASA interview on the phone that evening. "It means that when I finish my degree I'm either going two hundred miles up or six thousand feet down," I replied breezily. After a long silence, she spoke again, this time with a very noticeable strain to her voice: "Isn't there anything exciting on the surface?"

She called back the next day (a rarity) to retract the comment, saying that her mother's support would have meant the world to her if she had been given such an opportunity, and that I had hers for both of the prospects ahead of me. Her sudden and untimely death almost exactly two years later would spare her the anxiety of watching me undertake either adventure and make me treasure all the more the pledge of support she gave me that day.

✷

"Tell us about yourself. Start with high school." End of instructions.

With that, George Abbey, the head of shuttle Flight Crew Operations, began the session that was both my very first job interview and by far the most momentous one of my career.

NASA had already determined that I was fully qualified for the job. That is why I was among the 200 finalists, winnowed out from over 8,700 applicants, who they brought to the Johnson Space Center in Houston for a closer look throughout the summer and fall of 1977. Over the course of our weeklong visit, the twenty of us in this batch of candidates would endure a dizzying array of physical and psychological tests. The medical team took the phrase "closer look" quite literally. Privacy and dignity went out the window as they examined body parts we hardly knew we had from angles we never thought possible. The psychologists probed our mental health via a pair of interviews—conducted in classic good cop/bad cop fashion—and a stint of isolation in a spherical spacesuit that NASA at the time envisioned using to rescue a stranded shuttle crew. All the data, clinical notes, and general observations collected about us would determine whether or not we were physically and mentally qualified—whether we stayed in the running or were removed from contention. We could only guess exactly who would render this judgment and on what grounds, but the bottom line was simple: unless one of the medics pulled my name off the roster, my chances of becoming an astronaut rode on the interview that had just begun.

The task of the fourteen experienced space professionals sitting around the conference table in front of me—four of them astronauts whose daring exploits I had followed avidly in years past—was to take the measure of the person behind the professional credentials.[4] How did she react when facing a high-stakes situation with little or no guidance? Could she distill a complex subject down to its essence and present it clearly? Did she get so enthralled with her own words that she droned on *ad nauseam* ("stuck in transmit," in astronaut parlance), or did she demonstrate situational awareness by reading and reacting to her audience? This deliberately vague, open-ended question—"Tell us about yourself"—was designed to reveal all of that and more.

The interview was actually the finale of a two-part test. Part one had been a no-notice assignment given at the orientation session held the Sunday evening of our arrival: write a brief essay on why you want to be an astronaut. No more than one page. Due tomorrow morning. End of instructions. Mine, handwritten and no doubt on the cheap stationery from my hotel room, read as follows:

> The space program seems to me to offer an exciting challenge to earth and planetary scientists: to come to a new and deeper understanding of our native planet and the cosmos. From a personal point of view, my interest in being an astronaut reflects a natural expansion of interests that always have been interdisciplinary in nature and global in scope. Participation in this program would provide an unequalled opportunity to increase my own understanding in these areas while, and this is very important to me, benefitting other scientists in many fields and mankind in general. There is undoubtedly much more drudgery and lead-in work associated with a space mission than with an oceanographic one, but my experience with the latter has given me an appreciation for the value of dog-work [*sic*] that is well done. I would like to contribute what I can to what NASA is doing in the shuttle program.

George's question was the kickoff to part two. I began with a high-level summary of my life's path to that point: my early fascination with maps and all facets of geography. The enchantment with the grand adventures of the early space program and deep-sea exploration in my preteen years. My flair for foreign languages, and the desire for adventure that propelled me through high school. The discovery, during freshman year at college, that the earth sciences were a more promising path than languages toward the life of creativity and adventure I had longed for. Finally, how I loved going out to sea on oceanographic cruises and the parallels I saw between these expeditions and the shuttle missions NASA was preparing for. Then I stopped, putting the ball back in their court for another question.

My only other vivid memory of that hour-and-a-half is an exchange with George near the end of the interview. This was triggered by what I now know is the classic job interview prompt to recount an experience that illustrates how you recovered from a big setback or overcame a serious challenge amid difficult circumstances. I chose an episode from the previous summer's research cruise, which had been beset by problems. This expedition was my last opportunity to plug key holes in my maps; I had to have the data to complete my degree. The small vessel we had chartered for the voyage lacked the winches we would normally use to deploy and retrieve our heavy scientific gear, so we had to do this by hand. Late one night, near the end of the cruise, a sensor in the 1,500-foot-long cable we were towing behind us began to leak for the umpteenth time. Together with my major professor, who was chief scientist on the expedition, I gathered a couple of deck hands and we hauled the cable in over the ship's stern. Since our little vessel also lacked a useful aft deck, we had to zigzag the cable up and down the ship's central passageway to work on it. It was well after midnight by the time we finally found the leaky sensor, and my professor had had enough. "Fix the damned thing and get back on the line," he barked as he stormed off to bed.

George broke in at this point and asked, "So, what did you do then?" "We fixed the sensor, streamed the gear, and picked up the survey line," I replied, mentally adding a snarky "of course." His follow-up question seemed equally absurd: "And then you went to bed?" Again swallowing my shock ("Are you out of your mind? Of course not!"), I described how I stayed on duty through the remainder of that watch period to make sure the fix held, woke the oncoming watch team, briefed them thoroughly on the night's events, and only then went to bed. George muttered a quick "Thanks, I think that does it" and stood up, signaling the interview was over. I shook hands with each member of the panel and left the room, unsure whether the session had gone well or badly.

As I headed home to Halifax, I was certain that I could do the job of astronaut well and would love it, but I had no way of knowing where I stood among the two hundred people interviewed. All I could do was wait for the selection board's decision.

*

The ringing phone woke my roommates and me early on a cold Nova Scotia morning in the middle of January 1978. Nobody ever called us at that hour. Curiosity, tinged with the fear of some bad family news, drew us all toward the phone in the hallway of our apartment. The roommate with the phone at her ear was looking directly at me with very wide eyes. "It's somebody from NASA," she whispered, as she passed the handset to me. George Abbey was on the line. In the slow, deadpan voice for which he is famous, George asked if I was still interested in coming to work for NASA. I was amazed that he could make such a life-altering prospect sound so banal, as if he were offering me a job bagging groceries at the local market. I managed to say something along the lines of "Yes!" while my mind swirled with the implications and the unknowns of the very different career path that had just opened up before me.

A few weeks later, I found myself at the Johnson Space Center's Teague Auditorium in Houston, milling around backstage with thirty-four other lucky souls, as we waited to be introduced to the world as America's newest astronauts.

The people I was getting to know as new classmates included combat veterans, research scientists, engineers, and physicians. Like the astronauts who came before us, we were all intelligent, goal-oriented, competitive, adventurous, driven, and self-disciplined. The paths we had taken to the astronaut program were quite varied, ranging from university laboratories to military ready rooms, and hospital operating rooms to corporate engineering departments. Some of us were already proven leaders skilled at working in large organizations, while others were just starting their careers. For the three freshly minted twenty-six-year-old PhDs in the bunch—Steve Hawley, Sally Ride, and me—NASA astronaut would be our first full-time job.[5]

The shuttle marked a radical change, from the Mercury-Gemini-Apollo era of brief forays into space to a new era of living and working in orbit on a whole new scale. It was a versatile craft that could be equipped to serve as a temporary orbiting laboratory or to deploy, retrieve, repair, or refuel satellites. To some commentators, this was a comedown for NASA. The outfit known for boot prints on the moon and the dramatic save of Apollo 13 would now run a trucking service, shuttling people and cargo to low Earth orbit and back. I am sure none of us felt that way. We were excited to be in on the ground floor of something so novel and challenging. The thought that we were writing a new chapter of spaceflight history was very heady stuff for NASA's newest batch of astronauts.

3

THIRTY-FIVE NEW GUYS

As I walked across the stage to take my place among my classmates that February day in 1978, I felt a mixture of excitement about the grand new adventure ahead and amazement at the scale of the media circus surrounding me. The fact that new astronauts were joining NASA for the first time in nine years, plus the significance of the new group's arrival for the space shuttle program, was enough to make the moment newsworthy, but there was more. The frenzy that greeted us that day was heightened by the fact that our class included ten people who were unlike any previous astronauts: six women (Anna Fisher, Shannon Lucid, Judy Resnik, Sally Ride, Rhea Seddon, and me), three African Americans (Guion Bluford, Fred Gregory, and Ron McNair), and one Asian American (Ellison Onizuka). The media was buzzing about the novelty of the ten of us, especially us six women. NASA had promised that all thirty-five members of the new class would be available for interviews and photographs from the conclusion of the introduction ceremony until the end of the prime-time evening news hour, but it was quickly apparent that the media had little interest in any of the white male test-pilot astronauts in our class. Within an hour or so, the twenty-five who fit that description were on their way home or to the gym. The ten of us who were novelties began a barrage of national press interviews that would last well into the evening hours.

The only preparation we six women had for the media gauntlet ahead was a short coaching session with Dr. Carolyn Huntoon. Carolyn had been the lone woman on the panel that selected us all. A Louisiana native with a doctorate in physiology, she had come to the space center as a postdoctoral fellow in 1968 to study how astronauts' bodies adapt to spaceflight and develop protocols to maintain their health. She had risen through the scientific management ranks and was then the most senior woman at Johnson. We did not know it at the time, but Carolyn had already done much to smooth the way for the six of us—averting or defusing issues and countering the fears and biases revealed by questions about our clothing, romantic attachments, and even the new women's locker room in the astronaut gym. Our brief conversation before the interviews began launched a mentorship and friendship that lasts to this day.

Her advice boiled down to: start smart and stick together. The interviews we were about to give would produce the world's first significant impressions of us, America's first women astronauts. All the reporters were highly skilled interviewers, and every one of them was eager to get the personal tidbit that would generate the most buzz. Carolyn suggested we caucus about how we wanted to handle the very predictable questions we were sure to get about hair, makeup, and relationships. She encouraged us to be alert to the stereotypes our answers might play into and, if possible, arrive at a consensus about where we would hope to draw the line between our public and our personal lives. We went off to our first one-on-one national media interviews with Carolyn's words ringing in our ears.

Thirty minutes later, it was musical chairs time, as the six of us rotated to our second interviews. It just so happened that our paths through the media center's hallway crossed right by the women's restroom. "In here," someone said (Sally, I think), and into the restroom we went. None of us needed a bathroom break, but all of us were eager to compare notes on our first interviews. The restroom was a safe place, off limits to the all-male collection of reporters and NASA public affairs staff in the building. What was the *New York Times* guy interested in? How did you handle that? Who had ABC television next? A few minutes later, we were out of the restroom and on to our next interviews. And so it went throughout the afternoon and into the evening until the last interview was done.

Happily, NASA clamped a tight media lid down on us very quickly after that so we could focus on training. Although we were now officially members of the astronaut corps, we would soon learn that we were not full-fledged astronauts in NASA's eyes. We were merely astronaut candidates, or ASCANs for short. We would not be eligible for flight assignments until we completed a year-long basic training syllabus. Before we started that, however, we needed to choose a nickname for our class. We quickly settled on "Thirty-Five New Guys" or TFNG, which was both an accurate description of our group and a clever play on the military variant "the f****** new guy." Somewhere along the way, we coined the tag line "Ten interesting people and twenty-five standard white guys," to reflect the media's myopic and sensationalized view of our class (figure 3.1).

ASCAN training is like grad school for astronauts. The syllabus is designed both to give new astronauts a broad and solid foundation of knowledge and to forge bonds among the members of the class. It includes academics, flying qualifications, familiarization tours of all other NASA centers, and a series of seminars

FIGURE 3.1

The unofficial "Thirty-Five New Guys" (TFNG) class emblem. Source: Author's personal collection; artist unknown.

with veteran astronauts intended to transfer the lessons learned on previous missions. The nation's leading experts were brought in to give us a crash course in every technical subject related to flying in space, from orbital mechanics to spaceflight medicine, oceanography to aerodynamics and space physics to principles of spacecraft design. We were fitted for flight suits, helmets, and parachutes and put through Air Force survival courses to prepare us to fly NASA's sleek and agile T-38 jet trainers. Legendary Mercury, Gemini, and Apollo astronauts—the people I had grown up reading about and watching on television!—came to the space center to share the lessons they had learned as they prepared for and flew their historic missions and to teach us by example how to be effective in the many roles we would play as astronauts. I went home every day thinking, "I can't believe I get to do such amazing things!"

✷

The first shuttle flight seemed to be both approaching rapidly and slipping constantly throughout our ASCAN year. I followed the progress and setbacks mainly via the reports given at each Monday's All Astronauts meeting by the more senior astronauts who were working in the various technical areas involved, such as software verification, checklist development, flight controls, simulation and training, and cockpit displays. Spare moments at the gym or our favorite happy hour watering hole were often filled with talk about the cargo manifest, which laid out the sequence of flights scheduled for the future. Each of the shuttle's early missions was designed to validate and showcase the capabilities of NASA's new space truck. The remote manipulator arm would go through shakedown tests on the second test flight and be used to deploy and retrieve a small free-flying satellite a few flights later. An early spacewalk would test the new shuttle-era spacesuit, demonstrate the ability to conduct EVAs, and pave the way for satellite repairs and a refueling demonstration several flights later. Scientific operations would start with small clusters of instruments and then expand quickly to include Spacelab missions that filled the entire cargo bay with a pressurized laboratory module and large pallets of scientific equipment.

Veteran astronauts had already been named as the crews for the first four shuttle flights, but the assignment roster beyond that was blank. From about the fifth flight onward, the meat of each shuttle mission would be some mix of deploying and retrieving satellites, operating the shuttle's manipulator arm, running scientific instruments, conducting experiments, and doing spacewalks—all tasks that would fall to mission specialists like me. Our mouths watered at the thought of being the lucky one who would get to snatch a satellite out of the sky with the robotic arm, float out into the payload bay on a spacewalk to repair an ailing satellite, or—better still!—clamp on the new jetpack known as the MMU (for Manned Maneuvering Unit) and fly off solo in the world's smallest manned spacecraft. We studied the flight schedule and pored over everything that might offer a clue to when our class would start to fly, who would go first, and who would get to do which of these incredibly cool things.

*

ASCAN year ended with a brief ceremony and our first jobs. Now that we had finished basic training, we were entitled to wear the shooting star pin that is the symbol of NASA astronauts (figure 3.2). Chief Astronaut John Young presented each of us with our silver pin at one of the weekly All Astros meetings. We would be given a gold one when we returned from our first flight. We also got our first real job, or "technical assignment" in astronaut-speak. Astronauts who are not training for a specific mission are assigned to work alongside the countless engineers and scientists on the ground who prepare the spacecraft and cargo for launch, design future missions, solve current engineering issues, or develop new capabilities. As rookies, we expected to rotate from one assignment to another roughly every fourteen months, learning the many different aspects of flight operations, mission planning, and engineering development. I thought of it as starting in the mail room and working my way up the ladder, a learning-by-doing way to understand the many strands of work that go into a

FIGURE 3.2

NASA astronaut pin. Source: NASA.

spaceflight and what it takes to bring them all together to make a mission successful.

The reasoning behind the particular technical assignment each of us got was a complete mystery to us, as would be the logic behind every other assignment in our astronaut careers, along with the process by which the decisions were made. Assignments were like a commander's order or a coach's decree: given without explanation and not up for debate. We could only speculate about the mix of factors that had been considered and what our first assignments might say, if they said anything at all, about how we had been judged so far or our future flight prospects. And speculate we did, sometimes obsessively. Everyone was eager to get in on the action of preparing for the first four shuttle flights, betting that our odds of an early flight assignment probably rose if we worked closely with the veteran astronauts flying these missions. My first assignment—helping to develop the systems management checklists for the shuttle's first test flight—was pretty good on both those counts. But after about nine months, I was reassigned to be a mission manager in NASA's high-altitude research aircraft program, based five miles up the road at Ellington Air Force Base.

I had mixed feelings about this change. On the one hand, I welcomed the escape from the endless meetings of the checklist world and the chance to fly aboard such a unique aircraft. NASA's WB-57F (figure 3.3) was a highly modified version of a British-designed jet that was almost as old as I was (the original Martin Canberra entered service with the United States Air Force when I was two years old).

On the other hand, I fretted that the assignment put me out in the boonies, far away from where the real action seemed to be. I consoled myself with the fact that I would get certified by the Air Force to fly in the same pressure suits that the four shuttle test-flight crews would wear. Surely that would be a good step toward

FIGURE 3.3

NASA WB-57F high-altitude research aircraft. Source: NASA.

wearing a spacesuit someday! Another and much more amusing tie to the shuttle's future flights came along soon after I earned my pressure suit qualification. Since I was the first woman the Air Force had certified to fly in a pressure suit, I was both the obvious and the only person who could conduct real-world operational tests on the female urine-control device that NASA engineers were developing for use in the shuttle spacesuit. Facing the prospect of being sealed in my pressure suit for up to twelve hours straight on long mission days, I was more than willing to do my part. Thankfully, previous laboratory tests had already taken the most cumbersome and uncomfortable gadgets out of contention, leaving only something called the Disposable Absorption Containment Trunk. The DACT, as it inevitably came to be called, was essentially an extra-absorbent adult diaper. It soon became standard equipment worn by even the most macho male astronauts for launch, reentry, and spacewalks.

✷

Sometime shortly after earning my pressure suit qualifications, I bumped into Bruce McCandless at the astronaut gym. Bruce was then the astronaut office's point person on the development of the shuttle's EVA capability. This meant he was often donning a spacesuit to work in the simulated zero-gravity environment of a neutral buoyancy training facility (essentially a very large swimming pool) to evaluate tools or experimental equipment proposed for future shuttle spacewalks. I felt a twinge of envy every time I learned that he had tapped one of my classmates to join him for one of these tests. They got a taste of what working in zero gravity would be like and a chance to show how well they could handle working in a real spacesuit. Trying my very best to maintain the astronaut-standard veneer of cool, I told Bruce about my pressure suit qualification and asked him to consider me for future neutral buoyancy runs. He obliged not long after, and so I found myself jetting over to Marshall with him in one of the sleek little T-38 jets we used for training and transportation.

The simulated spacewalk we would do that day was one step in the invention of an EVA that Jerry Ross and Jay Apt would perform in 1985. I use "invention" instead of "development" because the

EVA in question, like almost every spacewalk in the early shuttle days, involved the creation of new structures, tools, and methods to do things that had never been done before. This one was designed to test two methods that might be used to assemble the space station NASA hoped to build someday. Its specific goals were to test how long it would take spacewalking astronauts to become proficient at the various tasks, how quickly they would become fatigued, and what forces and torque they would apply to the structure during assembly. This information would tell the engineers how sturdy the pieces needed to be and allow them to predict how much one astronaut could get done during a single spacewalk. Our test that day was known as an “engineering run”: a run-through of part of the EVA, using prototype hardware and tools, designed to give the engineers astronaut feedback on the preliminary design of the structure and the assembly techniques. Everything had already been tested by engineers wearing scuba gear. Now they would find out how it worked for users wearing a spacesuit and learn what modifications would make it better.

Since the new shuttle-era spacesuit was still being developed, we would wear leftover Apollo-era suits for this test. I was given Pete Conrad’s old suit, because it came closest to matching my body measurements. I was thrilled to think I would make my first simulated spacewalk in the same suit once worn by a moonwalker, no matter that Pete only wore this one in the water tank. The thrill turned to pain once I started suiting up, however. Measured from crotch to shoulder, Pete was shorter than me by an inch or so. Squeezing my body into his slightly too small suit really took some doing. While the technicians went through their pre-dive checklist, I assessed the fit of the suit on every part of my body. The boots and gloves fit pretty well, but I could not move much without some part of the suit digging into my torso and boosting the pain level sharply. I wasn’t about to give up my first underwater run because of a crummy suit fit, however, so I stayed mum, hoping the suit would feel better when I got in the water, as the techs had promised.

Getting into the tank turned out to be the hardest part. At the Marshall facility, this involved walking from the rack that supported the weight of the suit during donning to the water's edge. That was much easier said than done when wearing over a hundred pounds of bulky gear that was a size too small. The metal ring holding my helmet dug into my collarbones when I tried to stand erect, sending bolts of pain through my body that took my breath away. I felt like a giant had grabbed me by the torso and was squeezing the life out of me. My suit technicians steadied me while I caught my breath and gauged whether or not I could go on with the test. Bending into a half-crouch eased the pain just enough for me to crab-walk over to the side of the tank. I was on the verge of vomiting by the time I got there and knew there was no way I could manage the stairs down into the water. Happily, the suit techs realized that as well and let me just flop into the tank. To my great relief, the suit squeeze and pain went away instantly when my body settled into its neutral buoyancy posture, which resembles the dead-man's float. They would no doubt return when it came time to get out of the tank, but that was hours away. It was time to go spacewalking!

The structure we were to build was a Tinker Toy–like tetrahedral truss (figure 3.4) made of six twelve-foot beams and four connecting nodes. The idea was for us to move along the structure as we worked, tethered safely but free-floating and working only with our hands. It was much harder than the Tinker-Toy analogy would suggest. The beams were about four inches in diameter over most of their length—too big for me to hold firmly with one gloved hand—and had a short two-inch diameter segment at each end. To connect a new beam to the structure, you had to get it aligned perfectly with a prong on the node and then slide a sleeve from the beam onto the prong until it locked into place. In orbit, you could get a beam into position by applying a small force to start it moving, waiting until it was aligned perfectly with the prong and then applying a counterforce to stop it. In the tank, it took constant force to move the beams against the resistance of the water. And since the skinny end of the beam was the only place that gave me a firm grip, I was always working at a huge mechanical disadvantage. Four hours of holding myself in place with one hand and swinging the beams with the other turned my forearms into spaghetti. I could barely unhook my tether to let the safety divers take me

FIGURE 3.4

Astronaut Jerry Ross works on the EASE structure during STS-61B in December 1985. Source: NASA.

back to the surface when we were done and needed both hands to pick up the can of soda the test conductor had waiting for me in the debriefing room.

Despite the initial pain of squeezing into Pete's suit and my worn-out hands, I loved every minute of the experience. I quickly learned when I could make my spacesuit do what I wanted it to do and when I needed to work the way it wanted me to. Controlling each limb and every muscle so I could maneuver smoothly and control the pieces of our Tinker-Toy structure was a fun athletic challenge, and I could see myself improving at it as the hours went by. Perhaps the most valuable lesson I took from the test was the importance of distinguishing things that were happening because I was in a water tank on Earth from things that would really happen in the microgravity of space. There is no such thing as a perfect simulation. Each one is deceptive in some way. In a business that depends so heavily on simulation-based training, both success and survival depend on spotting those deceptions and factoring them into your mental preparation. I would remind myself of this countless times in the years ahead.

I felt I had turned in quite a good performance for a first-timer, and Bruce and the Marshall neutral buoyancy test engineers seemed to agree. The test had really whetted my appetite for spacewalking, so I hoped it would mark me as a contender for an EVA assignment down the road. Of course, with the astronaut office's completely opaque approach to talent management, I had no way of knowing whether any appraisals of my debut performance would reach my bosses or affect my future assignments.

*

My hope of getting closer to the action on the early shuttle flights was fulfilled in early 1981. A few weeks before the launch of

Columbia on the inaugural space shuttle mission, STS-1, the astronaut office deputy called me in to tell me I had been tapped to provide "media support" for the mission. That meant serving as a mix of on-air presence and behind-the-scenes technical expert for one of the networks (ABC News, in my case). I was ecstatic, since this held the prospect of having a front-row seat at the Kennedy Space Center for the launch and at Edwards Air Force Base in California for landing. Eager to meet the broadcast team, I made my way quickly to the ABC News production trailer parked near the cluster of antennas that linked the Johnson Space Center to both the media world and the orbiting spacecraft.

There I learned that ABC had engaged Apollo 17 veteran Gene Cernan, the last man to walk on the moon, to be their on-camera expert, hoping to heighten the drama of the first shuttle launch. With that need filled, they assigned me, NASA's unflown rookie, to support the radio team. Vic Ratner and Bob Walker, the on-air personalities, and producer Shelly Lang welcomed me warmly, and I soon realized there were some distinct advantages to being on radio rather than television. We could dress casually and, when the moment of launch came, gawk and celebrate as much as we liked without embarrassing ourselves on national television.

The next few weeks were an intense blur of preparation. I went back over all my Shuttle 101 notes from ASCAN training, pored over the checklists and flight plan for the mission, and joined the production team's planning meetings. A few days before the launch, we flew down to Florida to start working out of the ABC broadcast booth at the Kennedy Space Center press site. Just three miles separated us from Launch Pad 39A, where *Columbia* stood awaiting liftoff.

Witnessing the maiden launch of the space shuttle transformed my sense of the great adventure I was involved in.[1] I saw the vapor plume at main engine start and watched the shuttle rise through the sky atop the column of fire from the solid rocket motors. A few moments later, the shock waves of the rocket blast enveloped me, shaking my entire body and thumping on my chest like fists. In that moment, "liftoff thrust" switched from being an abstract concept expressed in sterile numbers to a concrete, full-body experience that I understood viscerally. I was more than duly impressed by the power I had just felt, and more excited than fearful at the thought of riding on top of it someday.

I got yet another new perspective on shuttle launches and landings as the chase plane photographer for *Columbia*'s second flight in November 1981. For each of the first four shuttle landings, astronauts flying a T-38 jet rendezvoused with the returning space shuttle to answer two questions: were the shuttle's airspeed and altitude instruments working properly, and had the vehicle's thermal protection system, or heat shield, been damaged during flight? An aircraft determines its airspeed and altitude using probes that measure the air pressure outside the vehicle. On the shuttle, these so-called air data probes were recessed inside the fuselage to protect them from the extreme heat of reentry and deployed by the pilot during the final approach to the landing site. Accurate airspeed and altitude are important in any landing, but they are especially critical to pilots trying to land a big, heavy glider like the space shuttle. Would the probes provide accurate measurements after being subjected to the powerful vibrations of launch, the temperature swings and vacuum of orbit, and the heat of reentry? A chase plane flying alongside with well-calibrated air data sensors was the simplest way to check this. If the answer was no, the chase pilot could give the shuttle pilots accurate airspeed and altitude readouts as they glided toward the runway. The glass tiles covering the bottom side of the shuttle were the main concern with the thermal protection system. Over 20,000 of these protected the airframe and the crew from temperatures as high as 2,300°F. Photographs of the underside of the vehicle before touchdown would let the engineers identify any damage that had occurred during launch or in orbit, study how reentry heating had affected the damaged areas, and evaluate the safety margin of the design.

Getting those photographs was my job. My classmate Hoot Gibson, who was flying the plane, had the tougher task of intercepting the shuttle and flying on its wing during final approach. When STS-2 lifted off, Hoot and I were circling a few miles southwest of the launch pad at 40,000 feet, ready just in case something went seriously wrong within the first four minutes and the shuttle had to return to land at the Kennedy Space Center. We saw the flash of light as the engines ignited and watched the shuttle rise toward us atop a column of fire. In about a minute it zoomed past us on its way to outer space. Two days later, we were again circling at 40,000 feet, now over Edwards Air Force Base, as we waited to chase the shuttle down to its landing on the dry lakebed runway. Air Force radar operators steered us toward the intercept point as the shuttle streaked in from orbit. The shuttle descends so fast that we would have barely ninety seconds to get the photos, stabilize in formation on the starboard wing, and radio airspeed and altitude confirmation to the crew. It went by in a flash. Even though I only saw it through the viewfinder of my camera, a vivid image of *Columbia* flying right next to us against a backdrop of blue sky and tan California desert remains etched in my mind to this day, and the photograph I took as we flew alongside her (figure 3.5) hangs in my library.

My next assignment was a real plum: support crew at the Kennedy Space Center for the upcoming shuttle flights, along with fellow TFNGers Don Williams, Loren Shriver, and Steve Hawley. We nicknamed this role "Cape Crusader," a play on Batman's moniker "The Caped Crusader," to distinguish the subset of the support crew cadre assigned to work at the Kennedy

FIGURE 3.5

Photo taken by the author of the space shuttle *Columbia* on final approach to Edwards Air Force Base, November 14, 1981. Source: NASA.

Space Center from those in Houston. Getting a space shuttle and its cargo ready to launch involves dozens of intertwined strands of critical activities, hundreds of people, and countless meetings. This would be my first in-depth exposure to the complex organizational structures and procedures NASA used to pull all this together successfully. As support crew, we would cover all the prelaunch preparations for the flight crew back in Houston so that they could stay focused on training to carry out the mission. Serving as their eyes and ears, we would make sure that testing at the Cape covered the issues they were concerned about and that all relevant data and information got back to them directly or into the appropriate checklist. Cape Crusader was a coveted assignment because it provided hands-on experience with real flight hardware and operations, both of which were seen as good stepping stones to a flight assignment.

I had two roles in the preparations for STS-3, the third space shuttle mission: payload integration and prelaunch switch list. The first involved testing the pallet of scientific instruments that made up the mission's primary cargo. My job was to make sure that the test plans covered all the signals, indicators, and controls that Jack Lousma and Gordon Fullerton would rely on to operate the experiments in orbit, and to ensure that the data from the flight hardware matched the data in the crew's checklists. I would stand in for Jack and Gordo during the first round of testing, which would take place in a building dedicated to testing payloads as they arrived at Kennedy. On the second round, which would take place after the hardware had been installed in the shuttle, I would join them in the cockpit. After five years in the astronaut corps, I would finally be working with real space hardware and getting inside a space shuttle.

Space shuttle cockpits bristled with switches and circuit breakers; over 2,100 in all. Needless to say, it was imperative for every switch and circuit breaker to be in the correct position at liftoff. The only ironclad way to ensure that they were was to run a switch list, meaning to methodically check each and every one of them. This began about ten hours before liftoff during shuttle countdowns. At the insistence of the astronaut office, it was always conducted by astronauts—Cape Crusaders—rather than ground team technicians. Just before midnight on March 21, 1983, Loren Shriver and I clambered into *Columbia* to perform the switch list. It would turn into the worst night of my astronaut career.

The astronauts who had preceded us as Cape Crusaders for the first two shuttle launches had warned us about the switch list. Steady work progress and schedule predictability were not the launch team's strong suit in those days. The switch list procedure had run way over the allotted time during the first two countdowns. The Cape Crusader cadre had no spare bodies to relieve the astronauts who were in the cockpit when this happened, so they were left to struggle as best they could to remain alert as the hours dragged on. It was a setup for a serious mistake, warned Ellison Onizuka; someone was sure to get bit unless it was fixed.

Sure enough, the switch list ran long on STS-3. As the team in the launch control center worked to resolve each hiccup, Loren and I, along with the ground technician who was with us, chatted and moved around the cramped cockpit as much as we could to keep ourselves alert. We had all slumped down onto the floorboards when launch control broke a long delay with a radio call to me. I gathered myself up, grasped the control stick at the pilot's seat, and pushed the button on top of it to respond.

Suddenly, *Columbia*'s cockpit lit up like a Christmas tree. Warning lights flashed brightly and alarms clanged loudly all around me. I was mortified when I realized what was going on. Onizuka's prophecy had come true, and I was the one bitten. I had just flipped the shuttle's computers over to the emergency backup flight control mode. Drowsiness had caused my airplane habits to take over, guiding my thumb to the button on the top of the control stick that activated the microphone in our T-38 jets and every light aircraft I had ever flown. I was vaguely aware even as I pushed it that the button felt much stiffer than a typical mic button, but the deed was done.

It was a serious mistake on a huge public stage in prime time.

I swept aside grim thoughts about what I had just done to my budding space career to deal with a more immediate problem: making sure my screwup did not jeopardize the launch. It was clear from the anxious chatter on the voice loops that the launch team feared this had been an "uncommanded engage," meaning that some gremlin in the computer systems had caused or allowed the backup system to take over on its own, without a command from the cockpit. If there was even the slightest suspicion of this, the shuttle would no doubt be grounded for exhaustive trouble-shooting. I got on the voice loop to the launch director and confessed: I was the reason the backup system had taken over; no gremlins lurked in the software.

The only silver lining to my dramatic "oopsie" was that it was completely reversible. After some additional discussion within the control center, the team began the procedure of returning the computers to their normal state, in preparation for continuing the countdown. I stopped by the launch control center when we were finally done, to give the launch director and computer team a complete account of the night's events and answer their questions. The shuttle launched as planned later that morning.

I seriously considered skipping the traditional postlaunch party. Neither the fact that my goof had been due to fatigue rather than stupidity or carelessness nor that it had not affected the launch lessened my embarrassment and shame. Feeling like the team goat, I dreaded the thought of facing my colleagues and enduring the ribbing that was surely coming. I also knew better than to run from it, however, and that it would be best to fight fire with fire or, in this case, ribbing with humor. So, I gathered a few props, screwed up my courage, and headed for the party.

The launch team's gag gift for me was a gray metal box with two large red buttons on it, one labeled "This One" and the other "NOT This One." The launch director described it as a custom-built remedial training aid as he presented it to me. I accepted it meekly and apologized for marring the otherwise smooth countdown with a bonehead mistake. I then assured the team that I had already taken concrete steps to ensure I never did it again. With that, I brought my right hand out from behind my back, revealing a bundle of bandages that made it look like the offending thumb had been amputated. That got me a good laugh, but didn't make me feel any better.

The practice of debriefing accidents and errors thoroughly in order to learn from them is ingrained deeply in NASA's culture. I had done this with the launch team after we finished reconfiguring the computers and completed the switch list, and expected there would be an astronaut office debrief when we got back to Houston. If there ever was one, it did not include me. Only one person in Houston ever said anything to me about the event. My classmate Dick Scobee, seeing that I was still beating myself up about it some weeks later, assured me that it was no big deal but that I should expect to be teased about it for a long time. I knew he was right about the second point, but it would take me a long time to believe him on the first one.

✷

The fourth shuttle flight was the last of the so-called Orbital Flight Test missions in the shuttle program. Essentially shakedown flights, each one carried just two astronauts and a modest cargo. Upon *Columbia*'s landing at Edwards Air Force Base, scheduled for July 4, 1982, the shuttle would be declared "operational" and begin carrying crews of five to seven astronauts and more complex payloads. The newest space shuttle, *Challenger*, would depart Edwards that same day, mounted on the back of the highly modified 747 jetliner that would ferry it to the Kennedy Space Center in Florida to begin preparations for its maiden flight in April 1983. Shortly before launch, we learned that President Ronald Reagan and First Lady Nancy intended to be at the landing to help celebrate this trio of grand events.

This final Orbital Flight Test was also the shuttle's first mission for the Department of Defense. Since the payload was classified, the payload integration role went to Army astronaut Woody Spring, the newest Cape Crusader. Most of my time for the ensuing twelve months would be spent on the team that was getting Kennedy ready to handle a new rocket booster that was designed to transfer satellites from the shuttle's orbit up to geosynchronous altitude. The first satellite to ride this booster would itself be a first—the first of NASA's tracking and data relay satellites. Our team's job was to make sure Kennedy was ready to handle, test, and install this new booster by running a pathfinder vehicle—an exact replica of the booster, minus the explosive propellants—through every cargo handling facility, test stand, and checklist that the real vehicle would encounter between arrival at the center and liftoff.

This assignment did not relieve me entirely of STS-4 responsibilities. I was again paired with Loren Shriver for the prelaunch switch list. Immediately after launch, Loren and I would fly to Edwards to await the shuttle's landing and serve as what is known as the exchange crew. Our job there was to board the shuttle as soon as the hatch was opened and take over for Ken Mattingly and Hank Hartsfield, freeing them to greet the president and first lady, who would be waiting at the bottom of the stairs. Loren and I, meanwhile, would work with the ground crew to deactivate the shuttle's systems and prepare it to be towed off the runway.

Looking through *Columbia*'s forward flight deck windows as the 747 carrying *Challenger* climbed into the sky, I had no idea that I would ride that very spaceship into orbit just twenty-seven months later.

Loren and I returned to Florida after our STS-4 duties and dove right into launch preparations for STS-5, the mission slated to include the first planned shuttle spacewalk. I had let it be known that I wanted to be the Cape Crusader tasked with loading the EVA equipment aboard the shuttle and ensuring that the spacesuits checked out properly before launch. I got my wish, but things didn't work out as well for the STS-5 flight crew. Owing to a technical glitch on one of the spacesuits, Joe Allen and Bill Lenoir never left the shuttle's airlock. That bumped the first shuttle EVA to STS-6, the maiden flight of both *Challenger* and the booster rocket I had been working on since the spring of 1982. *Challenger* lifted off in the early afternoon of April 4, 1983. Three days later, Story Musgrave and Don Peterson inaugurated the new shuttle spacesuit on the first EVA of the shuttle program. My stint as a Cape Crusader ended the following day. I headed back to Houston to take up my new assignment as the office lead for all things EVA and technical adviser for several payloads that were lined up for future shuttle flights.

✷

Anticipation began rising in early 1982 that the TFNG class would soon start to get flight assignments. Comparing bets about who was most likely to get tapped first became a common topic at happy hours, in quiet moments around the office, at the gym, or on cross-country flights. We were all eager to fly but knew that the majority of us would not be on the first slate of names. That would be a tough pill to swallow for a group of people accustomed to finishing first. Someone came up with five rules of spaceflight to lend a bit of perspective to the disappointment that most of us would soon face:

1. There's no such thing as a bad spaceflight.
2. Some flights are better than others.
3. Flying sooner is better than flying later.
4. Longer flights are better than shorter.
5. High-inclination is better than low.[2]
6. When in doubt, refer to Rule 1.

Even greater suspense surrounded the question of who would become the first American woman to fly in space. I, for one, was confident that any one of the six of us would be a perfectly fine choice in terms of technical competency, but suspected that judgments about media appeal and personality were likely be factors in the selection. Not being a cover girl type, I reckoned that the first of those factors probably put me at a disadvantage. Of the remaining women, I figured the second factor strongly favored Sally, who was by far the most assertive and competitive personality among us—the kind of friend you might grab a beer with, but not one to bare your soul to. Sally also seemed to be in the fast lane, having worked with astronaut office superstars Joe Engle and Dick Truly on STS-2 and clearly holding a solid spot on George Abbey's list of favorites. I thought my assessment was eminently logical, but rather disheartening.

The first TFNG assignments were announced at an All Astros meeting in mid-April 1982. Four of our number, Sally among them, would fly on STS-7, and three more on STS-8. In keeping with astronaut office culture, both the lucky ones selected and the many left behind received the news with a game face. Those of us not named to one of the crews swallowed our disappointment and congratulated our classmates, while the lucky souls tried to keep their glee in check until day's end, when they would meet at a local watering hole to celebrate their good fortune.

✷

June 18, 1983: STS-7 launch day. This would be the first shuttle launch I did not watch from the Cape. I was in California instead, invited by the University of California at San Diego to give a commencement address the following day. I had seen two good reasons to accept the university's invitation, one practical and one personal. On the practical side, the trip would let me stop by Scripps Institution of Oceanography to complete my open-water scuba qualification, something I had been trying to fit into my schedule for many months. On the personal side, it would spare me the challenge of keeping up a happy face as I watched the flight I had hoped might be mine leave the pad. At the time of liftoff, I would be preparing for my checkout dive instead of sitting glumly in the small astronaut office conference room back in Houston.

✷

The large crowd gathered at Kennedy Space Center to welcome America's first female astronaut back to Earth on June 24 was a very VIP bunch. Everyone there had received a personal invitation from either the White House, the NASA administrator, or Sally herself. They would have the twin priviliges of witnessing the first space shuttle landing in Florida and being the first to meet the crew after landing. Despite the predawn hour, everyone was bubbling with the mix of excitement and suspense that always surrounded shuttle landings.

Back in Houston, I was following the landing preparations via the mission audio and video loops, which were piped into the astronaut office conference room. The weather at the Cape had been rotten all night long and was showing no sign of getting better. After hearing yet another gloomy forecast, I left the room to freshen up my coffee. No decision had been announced yet, but everyone in our conference room was sure that *Challenger* would be redirected to land at Edwards Air Force Base. Deputy office chief P. J. Weitz intercepted me as soon as I stepped into the hallway. "You and I are going to Florida to entertain the VIPs," he said. "Meet me at the airfield as soon as you can."

Thirty minutes later, we were jetting across the Gulf of Mexico in a T-38. It would take another couple of hours for *Challenger*'s orbital track to line up for a landing at Edwards. We had to get to the Cape before that to entertain the VIP crowd while they waited for the shuttle to land in what was, to them, the wrong place. Perhaps they already knew that they would not have the thrill of being the first to greet America's now most-famous astronaut after all. I wondered how they would feel when the two of us showed up as their consolation prize.

I was really not looking forward to this appearance. I suspect P. J. felt the same, but at least he was a veteran astronaut who could speak firsthand about flying both the Apollo command module and the space shuttle. I was still a rookie. Sending me to meet this crowd was like sending a rookie Olympian to stand in for a world-famous gold medalist. All the cachet and curiosity that excited the crowd stemmed from an extraordinary peak experience that the stand-in knew nothing about.

The auditorium door opened on a huge crowd buzzing with excitement. I was instantly very glad that Sally was out in California. She would have a few hours alone with her crewmates to bask in the afterglow of their successful mission, something anyone would treasure after such an extraordinary experience. Plus, the trip back to Houston would give her some quiet time to absorb the impact the flight would have on her life and prepare herself for the media frenzy that awaited. She would have gotten none of that if the morning had gone as originally planned and *Challenger* had landed on the Kennedy Space Center runway. Instead, she would have been the one who was about to be swept into the whirlwind of energy that I saw, heard, and felt in the room around me.

The second thought that crossed my mind was "If this is what you get for going first, she can have it!"

✷

My next technical assignment, in July 1983, put me into what we called the Mission Development group. This rotating cadre of mission specialists monitored and took part in the development of payloads—scientific instruments, EVA equipment, deployable satellites, and so on—slated for shuttle flights that were too far downstream in the schedule to have a crew assigned to them yet. Our task was to inject the astronaut's operational perspective into the design of the hardware and operating procedures and ensure that any issues relevant to crew health, safety, or efficiency were spotted early and dealt with effectively. In one sense, we were technical consultants working for the payload teams, but we were consultants with a clear (though unwritten) mandate to work primarily on behalf of the eventual flight crew.

Sometimes the payload we were assigned to track was in the very early stages of development, still many years away from its flight date. The two payloads assigned to me in April 1983 were scheduled to fly sometime in 1984. One was a suite of earth remote sensing instruments called OSTA-3, after the NASA science office that funded it (Office of Space and Terrestrial Applications), and the other was an orbital refueling experiment that went by the obvious acronym of ORS (Orbital Refueling System). With our class now being named to crews, I dared to hope I might get to fly the mission carrying these two payloads, instead of handing them off to someone else.

I didn't appreciate it fully at the time, but the refueling system assignment put me squarely into the action on one of NASA's highest priorities in the early shuttle era: convincing would-be customers that the shuttle was more than just a space truck that could deliver satellites to orbit. To attract more users, including international customers, the agency wanted to demonstrate that the shuttle could also be used to retrieve, repair, and refuel satellites, including ones that had not been designed from the outset with this in mind. The Orbital Refueling System experiment was meant to provide solid proof of the third "R" in this triad.

Someone far above my pay grade had decided that the refueling demonstration would not be convincing unless it mimicked a real satellite very closely and involved transferring real, explosive propellant instead of a much safer liquid like water. The Landsat-4 satellite had been identified as a candidate for future on-orbit refueling, so the engineers in Houston made the test hardware an exact replica of that propellant system, right down to the same valve caps, seals, and safety wire installed on it before launch.[3] This was all to provide a good test of the reach, visibility, and dexterity factors an astronaut would have to deal with to refuel the real Landsat-4 satellite.

The live propellant was the toughest part. Hydrazine, the most common propellant used to maneuver satellites in orbit, is nasty stuff, both highly explosive and extremely toxic. People working with it on the ground wear high-tech full-body protective gear to make sure they don't inhale any fumes or get any liquid on their skin. If there's a leak, they can flee and wash down with water for good measure before getting out of their protective gear. Spacewalking astronauts would have none of those options. The safety experts had decreed that the refueling experiment's design had to provide two physical barriers between the spacewalkers and the hydrazine at all times. The challenge facing the ORS engineers was how to design a set of tools that would meet this demand while simultaneously allowing the astronauts to remove the seals, open the valve, and insert the new fuel line. Imagine trying to fill the gas tank on your car without ever touching the fuel lid or gas cap.

They were still looking for a solution when I began working on the project in mid-1983, along with Dave Leestma, who was then just another guy in the Mission Development group but would soon become my crewmate. This made the refueling experiment a special opportunity for me. Most of my previous jobs had been in operations, where the challenge was to master the use of existing tools and equipment, such as the instruments aboard the high-altitude research aircraft, the experiments being loaded into the shuttle's cargo bay, or the controls on my spacesuit. Even the Tinker-Toy neutral buoyancy test I did with Bruce had involved fairly mature hardware. I was excited to be getting involved in early-stage design with the refueling system tools and knew the experience would stand me in good stead for future EVA work.

"Getting involved" did not mean becoming a full-time member of the design team. Astronauts are meant to be generalists and system operators, not designers. Moreover, as one of NASA's scarcest resources, we astronauts were always in demand on multiple fronts. The refueling system was just one of several projects that Dave and I had to split our time among, but it was a fascinating one that got high priority because of its importance to the shuttle program. We joined the engineers' brainstorming sessions and added our ideas to the mix as they tried to come up with a way to solve the two-barrier problem. The starting point was obvious: the Landsat fuel valve had threads on it that first held the fuel nozzle when the tank was filled on the ground and then held the protective dust cap the satellite launched with. They envisioned a cylindrical housing that could screw onto those threads, with a second set of threads on the opposite end to hold each tool and the new fuel line. The engineers had not yet come up with the arrangement of seals or valves inside the housing that would both allow access to the fuel valve and satisfy the safety criteria. I think it was Dave who suggested in one of our creative sessions that they build a ball valve into the housing. Close this new valve, and it became another barrier. Open it, and our tools could reach through to the Landsat fuel valve. We could operate it by hand, with a simple lever like the one on a household water line. By the next time we met, the engineers had produced a design drawing of the new housing concept and identified a ball valve with the right dimensions and specifications that was made of a material which was compatible with hydrazine. More problems would crop up during fabrication and testing of the flight hardware, but the ball valve design proved successful in the end.

My hope that the earth science and refueling system payloads would be my stepping stones to orbit was realized in September 1983, when George Abbey called me into his office to tell me that I was assigned to the mission carrying these payloads. Our flight would be designated as STS-41G, in accordance with the new numbering sequence NASA Headquarters had adopted. (Popular rumor had it that a superstitious NASA administrator had directed the shift from numbered flights to this clumsier designation to avoid ever having an STS-13.[4] Ironically, STS-41G would turn out to be the thirteenth space shuttle flight.) I would be the crew lead on the earth science experiments, with Dave backing me up. Dave would take over the lead on the refueling system experiment, and we would do the satellite refueling EVA together. Our third mission specialist, Sally Ride, veteran of the seventh space shuttle flight that past June, would serve as flight engineer and also operate the shuttle's robotic arm to deploy the solar radiation monitoring satellite we were delivering to orbit. Commanding the mission would be STS-1 veteran Bob Crippen. Crip, as everyone calls him, was then preparing to command the mission that was to rendezvous with the disabled Solar Maximum spacecraft (a.k.a. Solar Max) and attempt the first ever on-orbit satellite repair. Rounding out the crew would be another TFNG classmate, Navy fighter pilot Jon McBride. We would remain a crew of five for just nine months. NASA would then add Royal Canadian Navy officer Marc Garneau and Australian oceanographer Paul Scully-Power as "payload specialists," the label the agency gave to anyone aboard a shuttle who was not a career NASA astronaut.

✷

For the space shuttle program, 1984 was both a tumultuous time and a banner year, with a series of missions designed to prove the shuttle's versatility and dramatic events that laid important groundwork for Hubble.

The year got off to a spectacular start with the first flights of the Manned Maneuvering Unit, a backpack designed to let EVA astronauts fly away from the shuttle to repair or retrieve satellites. The privilege of making the first test flight fell to Navy pilot and STS-41B mission specialist Bruce McCandless, one of the unit's codesigners. Hoot Gibson's photograph of Bruce as the world's first independent human satellite, flying alone above the blue arc of Earth, became one of the iconic images of the shuttle era. In less dramatic but equally important work, Bruce and his EVA partner Bob Stewart also tested some of the tools that had been developed to repair the ailing Solar Max satellite and a prototype fluid coupling that would pave the way for our orbital refueling experiment.

The drama and success of the spacewalks on STS-41B overshadowed two failures that would soon affect my own fortunes. Earlier in the mission, the crew had deployed two communications satellites for commercial customers. NASA hoped that commercial

FIGURE 3.6

Astronaut Bruce McCandless on the first test flight of the Manned Maneuvering Unit, February 7, 1984. Source: NASA.

operations like these would become the shuttle's bread and butter, and accorded them higher scheduling priority than the agency's own scientific missions. Each satellite was mounted on a rocket motor that was to propel it to its assigned orbit. Both motors had failed, leaving the two satellites stranded in useless orbits and putting a hold on every downstream flight that was scheduled to use a similar motor. As the engineers raced to figure out what had gone wrong and how to fix it, NASA juggled the flight manifest. Crews who found themselves moving up in the sequence, like we did, rejoiced silently, while those whose flights slipped or fell into limbo, with launch dates listed as "To Be Determined," cursed the fates. NASA would send a crew to retrieve the two stranded satellites before the year ended. That mission would succeed, but it would also remind us of the perils of designing satellite servicing equipment solely on the basis of engineering drawings.

The first cautionary lesson in this latter vein was learned in April 1984, on the world's first satellite repair mission, STS-41C. Failures in the main electronics module and attitude control system had crippled NASA's Solar Max satellite back in 1980. NASA saw the crippled bird as a golden opportunity to demonstrate the shuttle's capabilities. As James A. Abrahamson, then NASA's Associate Administrator for Spaceflight, put it in a memo to the Johnson and Goddard center directors, the on-orbit repair would be "a magnificent demonstration of the Shuttle, our astronauts and NASA's capability and foresight." Abrahamson went on to urge that "we must use this mission to help convince the international space community that the revolution in space is *here* and *now* and to persuade them to take advantage of orbital servicing, repair and assembly in the future" (emphasis his).[5]

The space sciences group at NASA's Goddard Space Flight Center had built Solar Max using a modularized, maintenance-friendly design they had developed in the 1970s. The space shuttle was on the drawing boards at the time, so they had added one of the specialized fixtures that would allow the shuttle's robotic arm to grapple it as a precautionary measure. The game plan was that Commander Bob Crippen would bring the shuttle close enough to Solar Max for astronaut George "Pinky" Nelson to fly over to it with an MMU and grapple it with a specially designed tool. The first EVA went smoothly, with real-time television scenes that seemed straight out of Buck Rogers, until Pinky tried to grapple the satellite. The jaws of the tool did not close properly, owing to a small stud that stuck out near the grapple point. The grapple tool design had been based on engineering drawings that supposedly represented the "as built" state of Solar Max with 100 percent accuracy, but none of them showed this small stud. The spacecraft was captured successfully the following day, thanks to Crip's superb flying skills and T. J. Hart's masterful maneuvering of the shuttle's robotic arm. Pinky and spacewalking partner James "Ox" Van Hoften completed all the repairs successfully on the mission's second EVA.

Ground-bound astronauts usually gathered in a small astronaut office conference room to watch shuttle missions via the NASA television feed. If I wasn't in a training session for my own upcoming flight, I was in that conference room, watching events unfold with a mixture of excitement and worry, plus the inevitable touch of envy. The Solar Max repair mission was a complete success and a spectacular demonstration of what the space shuttle and its intrepid astronauts could do, but the near-miss with the grapple tool on the first EVA was too important a lesson to brush aside: a trivial spacecraft feature—in this case, the small stud near the grapple point—combined with a minor failure to update an engineering drawing could doom a satellite repair mission to failure.

*

The pace and intensity of training for STS-41G picked up very quickly after the Solar Max mission ended. We were just six months away from launch, a milestone known as "L minus six" in astronaut shorthand. We had been eager to reach this milestone for two reasons: Crip would finally join us full time as our commander, and we would turn all our technical assignment and public relations duties over to other astronauts. Honing our skills as a crew and training with the flight control team who would run mission control while we were in orbit would be our sole focus until launch. Our simulator sessions grew increasingly long, allowing us and our flight control team to "live" all the events of a full flight day. They also became much more challenging, with our instructors injecting mock failures to test our knowledge of our payloads and the space shuttle's operating systems. For me, going through all this for the first time, it was a crazy mixture of exhilarating and overwhelming—yet another round of "drinking from a fire hose" but now with the highest of stakes. I marveled at the ease with which Sally and Crip, our veterans, seemed to glide through it all, and hoped I might someday have it all together like they did.

EVA training was the exact opposite. I felt right in my element, thanks to my five years of pressure suit experience, a goodly number of neutral buoyancy tests, and the spacesuit checkout work I had done at the Cape. Dave had adapted quickly to working in the spacesuit, and together we soon mastered the basic choreography of our spacewalk. The few remaining technical issues on the refueling system were relatively simple ones, such as how to confirm that no hydrazine got into the airlock with us at the end of the EVA (find a device sensitive and fast enough to analyze the air on the spot) and what to do if it had (go back outside and allow sunlight to bake it out of the suit).

One minor issue required some last-minute inventiveness on my part. My role on the refueling operation was checklist minder and inspector. At one particularly delicate point in the process, the engineers wanted me to confirm that a very small seal had been removed successfully. The only way I could get close enough to see such a small detail was to position myself above the experiment, directly in front of Dave. The problem was that the top of the experiment was just a thin piece of metal with a thermal blanket on it. It was much too flimsy to hold an EVA handrail or withstand the maximum forces that an astronaut moving around in a 350-pound spacesuit might induce. Knowing I would need only fingertip forces to move into position, I proposed that we simply sew two loops of tether strapping to the fabric covering of the thermal blanket. These would be more than adequate for the task. Attaching them with breakaway stitches that would rip out if too much force was applied would prevent damage to the experiment housing. After extracting a promise that I would never grasp the straps with more than three fingers, the safety engineers agreed to install them for our next neutral buoyancy simulation. They worked exactly as expected, so the engineering team started the paperwork needed to document the design, produce the straps, and install them on the flight experiment.

*

We had the traditional final visit with our families on October 3, 1984, my thirty-third birthday. I couldn't imagine a better way to celebrate the occasion. Our support crew decided not to put any candles on my cake, reckoning that the two solid rockets we would ride to orbit two days later would more than compensate for their absence. They were absolutely right about that.

The final two days of the countdown were a strange mix of comfortingly familiar and utterly surreal for me. The events were all familiar because we had gone through each one of them during our countdown dress rehearsal three weeks earlier: weather briefing; update on shuttle and countdown status; medical exam; flight gear fit-check. We filled the time in between events autographing a bottomless stack of photographs for Kennedy Space Center employees and the public affairs office, going for a run, studying our checklists, or watching television. It felt surreal because this time the very familiar set of events was leading to the extraordinary experience of leaving the planet. Or so I hoped. Years of watching launches delayed and going through simulations in which nothing ever worked as planned had ingrained an "I'll believe it when I see it" attitude in me. I was trapped between longing desperately and not daring to hope, and all I could do about it was watch the time tick by.

A knock on my bedroom door at three in the morning woke me for what I hoped was my last day on Earth. That thought really brought home the immensity of what we were about to do. I had slept well and felt rested, focused, and ready. We all dressed in crew polo shirts for the ritual photo session of the crew eating their last earthbound breakfast before heading to the launch pad (nobody ever ate during these sessions). After a final weather briefing and countdown status update (all still GO!), I packed up all the earthly personal belongings I would not need in space to leave with our ground crew, who would have them waiting for me when we landed (or give them to my grieving family, if something went horribly wrong). Then I headed to the suit room for the final helmet and oxygen checks. A calm intensity pervaded the morning, giving it a reverent feel. Everyone around us hoped they were seeing us off on a grand space adventure and would welcome us back again in seven days. All of us knew there was an outside chance we might never return.

Each time I had gone onto the launch pad during my Cape Crusader years to prepare a shuttle for someone else's launch, I had wondered what it would feel like to go out when it was my turn. It felt exciting, magical, and still not quite believable. The seven of us jammed into the elevator, rode up 195 feet, and walked out onto the catwalk that led to the shuttle hatch. I looked down at the vapors swirling around *Challenger*'s tail and listened to the external tank creak and groan as the metal adjusted to the super-cold liquid oxygen and hydrogen that had been loaded while we slept. The shuttle seemed to be alive and as eager to fly as I was.

The seating arrangements in the cabin dictated that Sally and I would board the shuttle last. We waited our turn in the small chamber just outside the hatch, known as the white room. We were keenly aware that the cameras above our heads meant our every move was being monitored by the launch control center, and perhaps broadcast on national television as well. After a few minutes of idle chit-chat, we decided we ought to appear to be doing something more important than merely waiting. Watches are always synchronized before a big mission in the movies, so we decided to pretend we were synchronizing ours (figure 3.7). Happily, there were no microphones in the white room to hear us saying,

FIGURE 3.7

Kathryn Sullivan (left) and Sally Ride (right) pretend to synchronize their watches while waiting to board the space shuttle *Challenger* on October 5, 1984. Source: NASA.

"What do you think the news anchors are saying about us right now" or "Do you think we've stretched it out long enough?"

Challenger roared off the launch pad when the solid rockets lit at 7:03 a.m. on October 5, 1984 (figure 3.8). Eight and a half minutes later, we were over England. Sally giggled as she let her pencil and checklist float in midair, visual proof that we were now in zero

FIGURE 3.8

Space shuttle *Challenger* soars into space, October 5, 1984. Source: NASA.

gravity. Crip's routine "main engine cutoff" call to mission control was met by a very angry British voice on the radio, scolding us for using a frequency that the Royal Air Force had reserved for training missions in the south of England that day. Looking over Jon McBride's shoulder, I caught my first glimpse of Earth in the forward windows. It looked like a beach ball, with beautiful swirls of white cloud on the blue field of the ocean. "Wow," I blurted spontaneously, "look at that!"

Our flight plan did not include any time for sightseeing right after engine cutoff, however. We had to quickly shift the shuttle's operating systems over to their on-orbit modes and get to work. Before our already long day ended, the flight plan called for us to check out the shuttle's robotic arm, deploy the Earth Radiation Budget Satellite, power up the Orbital Refueling System experiment and earth science instruments, and unfold the large antenna for the OSTA-3 imaging radar. We managed to get all that done by day's end, but not without a couple of tense moments. One of the satellite's solar arrays failed to swing into the right position because of a cold hinge. Giving it some time to bake in the sun and a few shakes with the robotic arm (Sally's idea) took care of that.

Once we had separated from the satellite, it was time for me to unfold the antenna on our most important earth science instrument, the Shuttle Imaging Radar. This was a pretty complex mechanism. Two sets of motor-driven hinges allowed the thirty-five-foot-long (10.7 meters) antenna to fold together into a three-leaf stack about eleven feet (3.3 meters) long. A set of latches held the three layers together and another set secured the whole stack to a support frame. All the latches and motors had to work properly to deploy the antenna and turn on the radar and, more importantly, to fold it back up again so that the shuttle's cargo bay doors could be closed for reentry. Naturally, Crip hovered right nearby to monitor this critical operation.

Following the checklist prompts provided by Sally, my backup on the radar, I released the support frame latch and then the latch holding the leaves together. As soon as I commanded the first leaf to unfold, the antenna began to oscillate wildly, with the partially open leaf flapping up and down and the whole stack swinging left to right. Visions of the collapse of the Tacoma Narrows Bridge—a notorious example of catastrophic structural vibrations—flashed through my mind. Nothing like this had ever happened in any of our preflight testing. Not really knowing what to do, I acted instinctively and commanded the second leaf to unfold. To the great relief of everyone staring out the aft flight deck windows, this stopped the oscillations. Our heart rates returned quickly to normal, but there was a lot of nervous laughter at dinner that night as we reflected on the day's events and imagined how the antenna could have been damaged or ruined if the oscillations had not stopped.

The third problem that cropped up early in the flight had a more significant impact on our day-to-day scientific operations and our EVA plan. Every shuttle was equipped with a small antenna that was used to send high volumes of data through a geostationary satellite and down to the ground. About the size of a modern-day rooftop satellite television dish, this antenna was mounted on the starboard cargo bay sill, just aft of the crew compartment. Once the payload bay doors were open, we swung the antenna arm out over the side of the payload bay and released the latches that held the dish in place for launch and landing. This allowed the antenna dish to tilt and pivot as needed to stay locked on the relay satellite. Our highest-priority science payload, the Shuttle Imaging Radar, depended critically on this data-relay capability. About thirteen hours into our flight, the antenna dish started to rotate wildly, owing to a failure in the circuit that controlled its position. Mission control came up with a two-part workaround to deal with this problem. The first part was to freeze the antenna by removing power from the affected circuit. For the rest of the mission, we would maneuver the shuttle itself when we needed to point the antenna at the relay satellite. The maneuvers would cost the OSTA-3 experiments some observing time, but this plan preserved as much of the science as possible.

Our in-flight maintenance training and onboard toolkit prepared us well for handling this problem, but the shuttle was not designed to make a detailed maintenance task like this easy. The offending cable was attached to an electronics module that was buried behind ten storage lockers mounted on the back wall of the shuttle middeck. Sally and Jon, our primary in-flight maintenance team, began the tedious process of taking down the lockers while the antenna experts on the ground worked to verify exactly which pin inside what connecter needed to be removed and to confirm that there would be no unintended consequences. Mission control radioed the cable and pin details in due course, and, after some gymnastic moves by Sally to reach the cable, they got the job done. Sally and Jon were only too happy to let Dave and me have the honor of reinstalling all the lockers, and we were both happy to get a bit of hands-on maintenance experience.

This episode illustrates several noteworthy points about the roles of astronauts during the shuttle era. As Valerie Neal points out in her book about spaceflight during this time, we were the on-scene figures in everything, serving as pilot, operator, scientist, cook, medic, and maintainer.[6] The engineers in mission control had more information at their fingertips, thanks to telemetry from the shuttle and access to documentation, but knowledge and data alone do not fix anything. Only the astronauts onboard the spacecraft could do physical maintenance. As a result, the orbiting wrench-turner was sometimes the most celebrated and essential person on the entire flight team, at least temporarily.

The second part of the workaround involved our EVA. With the motors that moved the antenna dish out of commission, we needed another way to align the dish with the antenna arm so it could be locked into place for reentry. Positioning it manually during our EVA was clearly the answer, but this meant delaying the spacewalk until as late as possible in the flight. The flight plan updates we received the next morning slipped it from flight day five to flight day eight, the day before reentry. Spacewalks on the day before reentry were banned under the normal shuttle operating rules, because of the heightened workload and risk involved in both activities. The decision to schedule our EVA so late in the flight had to have been okayed at a very high level, and it clearly signaled how important this spacewalk was to both the orbital refueling demonstration and the safe return of the shuttle's antenna. Nonetheless, I worried that it somehow wouldn't happen at all. Fortunately, our flight plan kept me too busy over the following days to wallow in my anxiety.

I breathed a small sigh of relief when we started the initial preparations for our EVA on flight day seven. The first step was designed to get our bodies ready for the low-pressure, all-oxygen environment of our spacesuits. We lowered the pressure of the entire shuttle cabin from the 14.7 pounds per square inch (1,013 millibars) found at sea level to 10.2 pounds per square inch (703 millibars), about the atmospheric pressure at Copper Mountain, Colorado (base elevation 9,712 feet, or 2,960 m). Twenty-four hours of breathing air at this pressure would wash a lot of nitrogen out of our bloodstream, reducing the risk that Dave or I would get the bends when we got into our suits, where we would breathe pure oxygen at just 4.3 pounds per square inch (296 millibars). The two of us spent half of the day unpacking and testing all our EVA equipment to make sure all the spacesuit systems were working correctly. Everything looked good for our excursion the next day.

To my delight and relief, the flight plan updates we got the next morning confirmed that all was still GO for our spacewalk. It was EVA day at last! In just a few short hours, I'd be floating around in the cargo bay, stealing every glimpse I could of the Earth beneath me as I worked.

I couldn't wait to get outside, but there would be no hurrying through those final hours. Astronauts talk about "suiting up" for an EVA as if it were akin to putting on a hat or coat before heading out the door. In reality, "EVA prep," as it's called in NASA-speak, is the intense final preflight preparation of the body-shaped spacecraft that keeps us alive as we work in the vacuum of space outside our mothership. The EVA experts in mission control followed every step as Dave and I worked through our preparation, their scrutiny becoming harsher and harsher as we approached the critical moment of taking the airlock to vacuum. Any glitch they saw or concern that arose in mission control could bring everything to a screeching halt, and possibly even cancel the EVA altogether.

To the outside world, our EVA looked like a two-astronaut event, but it was really a much more complex production that involved our entire crew and mission control. Jon was our right-hand man for the EVA, helping us suit up, operating the airlock, and keeping track of our timeline and checklist. Sally had to switch off the radar payload before we left the airlock, so its transmitter wouldn't interfere with our spacesuit computers, and also monitor the ORS gauges while we were working on it. Crip had to maneuver the orbiter to an orientation that provided the right mix of sunlight in the cargo bay and communications coverage, in addition to orchestrating everything and worrying about the two of us outside. On top of all this, our three crewmates needed to capture key EVA moments on film for the IMAX movie *The Dream Is Alive* and open up the antenna electronics again to reconnect the locking circuit when I had positioned the failed communications antenna. It promised to be an intense day for all of us.

"*Challenger*, Houston: You are GO for EVA" were the sweetest words I had ever heard. My suit was humming along perfectly, and I was itching to get outside. Dave left the airlock first, with me right on his heels. We had trained for over a year to move quickly through the ORS work so that the radar experiment could start up again. This was no longer an issue, since the antenna problem had pushed our EVA to the end of the flight, but we set to work at our usual high tempo nonetheless. We hooked up our safety tethers, grabbed our tools from the toolbox at the bottom of the cargo bay, and started to make our way toward the tail of the orbiter, where the refueling experiment was mounted.

We had barely begun heading aft when Crip ordered us to stop. "Look around, guys. There are no scuba divers here. This is not the water tank," he said. "Take a moment to look at the Earth and appreciate where you are." I obediently pivoted my body away from the orbiter to look around. I was dangling from the handrail of a spaceship hundreds of miles above the Earth, which was gliding by below at 17,500 miles per hour (28,163 km/hour). A burst of clashing thoughts flashed through my mind. I was in this absolutely extraordinary place, keenly aware of how deadly the environment outside my suit was, yet being there seemed perfectly natural, and I felt utterly comfortable (figure 3.9).

After gawking at the Earth beneath us for what seemed like a long time but was probably just a minute or so, we got back to work. The fabric fingerholds I fought to put on the top of the experiment housing provided all the aid I needed to stay in position above the worksite and monitor each step as Dave performed it. The small seal the engineers had worried about came out without a problem. The only piece of refueling equipment that did not work exactly as expected was the two-piece clamshell bracket

FIGURE 3.9

Dave Leestma (left) and Kathryn Sullivan (right) at the Orbital Refueling System (ORS) worksite on STS-41G. The Shuttle Imaging Radar (SIR-B) antenna is on the left in the foreground. Source: NASA.

that had been added to hold the fuel line in place during reentry. Dave undid the clasps and rotated the two pieces out of the way at the start of the operation, as usual. They had always stayed firmly in place during our training, thanks to the tiny bit of friction on the hinges caused by gravity. Now they ricocheted slowly back and forth between the back of Dave's hand and the experiment housing, like small silver flags waving languidly in a gentle breeze. I was glad the only simulation inaccuracy we had failed to spot was such a trivial and amusing annoyance.

Repositioning the broken antenna so we could lock it into place for landing was my job. Dave and I had gone down the port side of the cargo bay to work on the refueling system. The antenna was on the opposite side. If I crossed over using the official EVA handrails at the very front of the cargo bay, I would not be able to reach the antenna dish. I needed to cross somewhere near the middle of the bay and then move forward along the starboard sill, and this meant improvising a route across the pallet carrying our earth science experiments. I had talked Crip through my planned route the night before, and mission control had finally blessed it. A bonus factor was that my transit would set up a perfect scene for us to add to *The Dream Is Alive*, with both spacewalkers in the foreground, and one right in the middle of the frame. As I made my way forward toward my crossing point, Jon was preparing the IMAX camera for the shot and Sally was pulling the last locker off the middeck wall so she would be ready on my command to reconnect the circuit that would latch the antenna dish back into its stowed-for-landing position.

The moment that is etched in my mind is from my traverse across the cargo bay. I was inching my way across the small shelf that held two of the earth science experiments, placing my hands carefully and keeping my boots pointed up toward the tail of the shuttle to make sure I did not kick one of the instruments. I felt like I was doing a handstand, and chuckled at what the scientists on the ground must be thinking as they watched. That shelf, like the top of the refueling experiment, was never meant to support a spacewalking astronaut. Jon's command to stop so he could film the IMAX scene gave me a moment to look around. As soon as I lowered my gaze from my hands, I felt like I was hanging from a tree limb and looking down at the ground. There, 140 miles (225 km) below me was South America and the Caribbean Sea. I watched in awe as the distinctive Maracaibo Peninsula of Venezuela slid between my boots.

All too soon, Jon said he was done, and I set off again toward the antenna. It moved to the stow position with the touch of a finger, and Sally drove the latches closed. Now we could safely bring it back to Earth for repair. We wrapped up our spacewalk with a final bit of improv that mission control did not know about. We had decided to film a scene that would mimic the "Kilroy Was Here" meme popularized during World War II (figure 3.10). Jon set up the IMAX camera to center shot on the four aft flight deck windows, two of which look aft into the cargo bay and two of which look straight up. All four would be empty and the shuttle cockpit would look deserted until Dave and I floated up and peered into the cabin. The soundtrack captured Sally saying, "I hear someone kicking the outside of the orbiter," just before Dave and I rose into view. We both laughed giddily through the entire scene, and I can be seen sneaking a few more glances at the Earth over my

FIGURE 3.10

One version of the "Kilroy Was Here" meme. Source: Wikimedia Commons.

left shoulder. When the camera stopped, we reluctantly we made our way back into the airlock. My first spacewalking adventure was over all too soon.

We spent our last full day in orbit packing up our spacesuits and scientific gear and taking a phone call from President Ronald Reagan, who wanted to congratulate the first American woman to walk in space, a.k.a. me. I savored every last glimpse of the Earth as I went about my chores, saddened by the thought that I would soon leave behind both weightlessness and the views of Earth that had so entranced me. The appointed time for our deorbit burn came all too soon. Some fifty minutes later, we came to a stop on the Kennedy Space Center runway. Looking around the lower deck of the shuttle, I struggled to convince myself that I had been floating effortlessly along the ceiling less than an hour before. My arms felt like they were made of heavy wet sand when I reached up to take off my helmet. I was back in the land where gravity rules. In under an hour, my first spaceflight changed from an immersive experience too stunning for words to something so distant that I wondered if I had really been there.

Worst of all, I was now at the very tail end of the long line of astronauts waiting for flight assignments. I would have happily jumped right back into one of our technical assignments, but suspected that the rest of my year would probably be filled with the series of technical debriefings that follows every flight and, given my newfound fame, countless speeches and public appearances.

*

I was right about what I was in for, but never imagined how extraordinary some of the events that awaited me would be. In early November, for example, I learned that I was to attend a White House dinner honoring the Grand Duke and Duchess of

Luxembourg, and that President Reagan intended to name me to serve on a commission he was establishing to recommend policies that would guide the nation's civilian space program for the next decade. The following week, a handsome Army major with the ornate presidential service aiguillette adorning his right shoulder escorted me through the formalities of a state dinner. I don't know whether the major or I was more shocked to discover that I was seated right next to the president, in the seat that protocol assigns to the third-highest-ranking guest. I was certainly floored to find myself exchanging dinner table pleasantries with the President of the United States.

Floored and a bit schizophrenic describes more accurately how I felt at his table. I had disliked Reagan as governor back when I was in high school and college and recalled how students at UC Santa Cruz had pelted him with eggs when he visited the campus the year before I arrived as a freshman. The same man beguiled me that evening with his charm and humor. He was an absolutely masterful host. Over the course of the dinner, he engaged each guest with a pithy question about their work or life that sparked a one-to-one conversation. After a few exchanges, he would make a remark to the entire table that shifted everyone back to chit-chatting with their neighbors. He did it all with warmth and graciousness, and no outwardly apparent effort.

The Army's Strolling Strings ensemble filled the room with elegant music as we dined on lobster en gelée, piccata of veal, and soufflé frangelica, and sipped fine California wines. I chatted with my nearest table neighbors, boxer Joe Frazier and golfer Patty Sheehan, whenever the president's attention was elsewhere. As dessert was served, I realized there was no electronic gear in the room and began to wonder if the traditional presidential toast to his foreign guests was to be skipped or not televised that evening. I was still musing on this when the two floor-to-ceiling columns on the wall opposite me swung open to reveal a battery of lights, and the dining room doors opened to admit a bustle of reporters and cameramen. The president was already on his feet, illuminated perfectly and ready to begin the toast. The lights, cameras, and reporters all disappeared just as quickly as soon as the president set down his glass.

Coffee was to be served in the China Room. I walked down the hallway on the arm of my escort thinking it was a bit odd for an entire room in the White House to be devoted to gifts from China and wondering what would be on display. I burst out laughing as we entered the room. The display cabinets lining the walls were filled with cups and saucers, dinner and salad plates, soup tureens and serving platters. Not gifts from China, but the tableware of previous presidents. I had one of those spine-tingling, time-has-collapsed moments as I sipped my coffee in front of a cabinet displaying President Grant's chinaware. Someone just like me (minus the spaceflight experience, of course) had eaten from those plates, in that very house, while our country was still recovering from the devastation of the Civil War.

✷

Another dramatic satellite servicing mission was underway while I was dining at the White House. The space shuttle *Discovery* had launched five days before on a mission to retrieve the two now-useless communications satellites that *Challenger* had deployed in February. The success of the Solar Maximum repair mission in April had prompted the insurance companies that were saddled with the failed satellites to lobby NASA to send a shuttle to retrieve them and bring them back to Earth for repair. Rick Hauck and his crew had been training nearly a year for the relatively simple mission of delivering two new communications satellites to orbit when NASA decided to add the retrievals to their flight plan. It was a bold addition, to put it mildly, and one that left the team with very little time to prepare for the most complex EVAs NASA had ever attempted. Wisely trying to temper expectations, Hauck stated flatly at their prelaunch press conference that it would be a near miracle if they pulled it off.

The retrieval plan read like a game of slow-motion hot potato with a multimillion-dollar satellite. Joe Allen would don an MMU, fly over to the Palapa B-2 satellite, and unfold an umbrella-like device known as the stinger inside the nozzle of the satellite's rocket motor. Once the stinger was locked firmly into place, the shuttle would fly over so that Anna Fisher could attach the shuttle's robotic arm to a grapple fixture on the side of the stinger. Anna would then move the satellite plus Joe over to spacewalker Dale Gardner. Working from a platform near the bottom of the cargo bay, Dale would attach a bracket across the top of the satellite that included yet another shuttle grapple fixture. Then Joe, still attached to the satellite via the stinger, would take control of the satellite again so that Anna could move the arm over to the newly installed grapple fixture. She would then position the satellite so that Joe and Dale could attach an adapter to the bottom of the satellite that would allow it to be berthed in the cargo bay and brought back to Earth.

Capture of the first satellite was scheduled on the day before I headed to Washington for the White House dinner. Eager to watch another Buck Rogers–style EVA, I claimed a prime spot in the small astronaut conference room where audio and video feeds from mission control ran continuously during every flight. Soon after the airlock opened, Joe set off in his MMU to capture the stranded satellite. We all fell silent as he made the final approach to grapple, thinking back to the problem Pinky had with Solar Max and hoping the 51A team had learned all the right lessons from that failure. Had we missed something else? Would the stinger work?

It did work, but the curse of relying on "as-built" engineering drawings returned soon after. The bracket designed to clamp onto the top of the satellite didn't fit. Just like Solar Max, the drawings the bracket designers relied on did not match the antenna structure on the top of the satellite. Now what? The NASA team again snatched success from the jaws of failure with some clever improvisation. While Anna held the satellite via the grapple fixture on the stinger, Joe undocked his MMU and moved to a work platform on the starboard rail of the cargo bay. Anna then gently placed the 1,200-pound

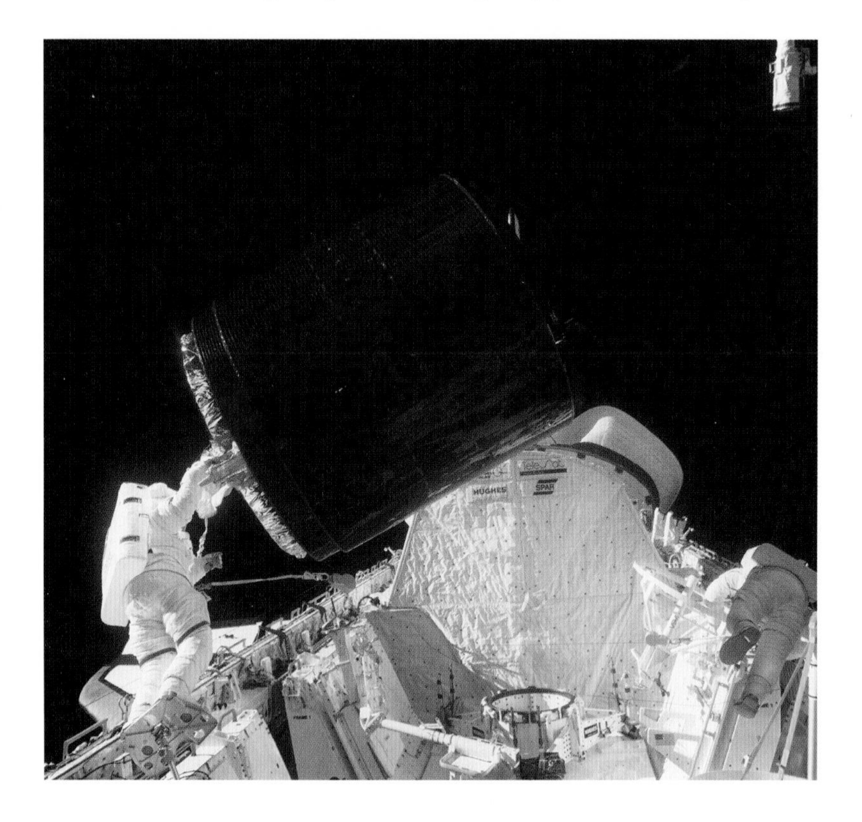

FIGURE 3.11

Joe Allen (at left) hand-carrying the 1,200-pound Palapa B-2 satellite. Source: NASA.

(544 kg) satellite into his gloved hands. Yes, Joe Allen hand-carried the Palapa B-2 satellite around the Earth for nearly two hours (figure 3.11), while Dale Gardner, who had shifted to a platform on the end of the shuttle's robotic arm, removed the stinger and installed the berthing adapter. After two hours of jaw-dropping scenes and very hard work, Joe and Dale manhandled the satellite into its berth. Two days later, they repeated this remarkable dance again and even more smoothly on the Westar 6 satellite, successfully completing the world's first satellite retrieval mission (figure 3.12).

FIGURE 3.12

Dale Gardner about to capture the ailing Westar 6 satellite. Source: NASA.

*

As 1984 drew to a close, the mood around the astronaut office was jubilant. In just seven months, we had proven the shuttle's ability to repair, refuel, and retrieve satellites. On every mission, the shuttle team—both flight crew and ground team—had adapted, improvised, and generally performed superbly. We were rightly proud of our expertise and ingenuity. My own life had changed radically over the preceding twelve months. I was now a flown astronaut and a spacewalker, with a historic first to my credit. With this new standing had come some amazing opportunities, like dinner at the White House and appointment to a presidential commission, plus a grueling slate of public appearances. Even so, beneath this whirl lurked that unpleasant fact that I was still at the end of a long line of astronauts, with no idea when I might return to space.

The best-ever belated Christmas present changed my fortunes for the better. George Abbey summoned me to his office a few weeks into the new year. He was assembling an all-veteran crew for the STS-61J mission, slated to launch in August 1986. The mission's primary objective was to deploy a top-priority NASA scientific satellite that had been in development for nearly two decades: the Hubble Space Telescope, as Spitzer's Large Space Telescope was now called. This immense spacecraft was supposed to be maintainable in orbit, but the tools and equipment needed to do this were nowhere near as ready as they should be with roughly eighteen months to go until launch. George needed to name the EVA and robotic arm mission specialists now, to help accelerate the development of the maintenance equipment and operational plans. The EVAers in particular had to get into the middle of things and make sure all the tools and equipment would really work as advertised. No maintenance was planned for the 61J mission, but the

EVA crew had to be ready to do a contingency spacewalk if mechanisms on the telescope's antennas or solar arrays failed during the deployment. Would I be interested in flying this mission, he asked? I left his office jubilant: I was out of the waiting line and back in the flight rotation.

4

"IT SHALL BE MAINTAINABLE"

Trying to do stellar observations from Earth is like trying to do birdwatching from the bottom of a lake.

—JAMES B. ODOM, HUBBLE PROGRAM MANAGER 1983–1990

Jim Odom's simile captures the raison d'être for the Hubble Space Telescope in a nutshell. Every astronomer since Galileo has surely longed to escape the clouds that block their view and the atmospheric turbulence that turns the pinpoint of light they want to see into a blurry, twinkling blob. They also want to study *all* of the light emitted by a star, not just the narrow, visible light portion of the spectrum that our atmosphere allows to reach a ground-based telescope.[1] The idea of putting a telescope into space would remain the stuff of science fiction until the advent of guided missiles, marked by the German V-2 rocket of World War II.[2]

When Princeton astronomer Lyman Spitzer sat down to write his paper about an orbiting observatory in 1946, he was able to draw on all the scientific and technological advances the American and German war efforts had produced.[3] Spitzer envisioned a telescope that would be at least ten times better than the ground-based telescopes of the day, able to see a mile-wide object on Mars, or a fifty-foot object (equivalent to one letter in California's famous "Hollywood" sign) on the Moon. This orbiting observatory, placed above the atmosphere and able to work twenty-four hours a day, could tackle familiar astronomical studies more rapidly and efficiently, but that was not his main selling point. The ambitious instrument Spitzer described would open altogether new frontiers in astronomy, allowing scientists to determine the positions and spectra of stars, understand the structure of galaxies and globular clusters, detect planets in other solar systems, and even determine the extent of the universe. His vision was nothing less than an observatory that would "revolutionize astronomical techniques and open up completely new vistas of astronomical research."[4]

I was stunned by the audacity of Spitzer's vision. The only object orbiting Earth at the time he wrote was the Moon. The basketball-sized beeper called Sputnik would not join it in orbit for another eleven years. Yet everything he imagined had come to pass. The Hubble Space Telescope, then nearing completion at the Lockheed Missiles and Space Company plant in Sunnyvale, California, was truly an extraordinary looking glass. It promised to be the biggest improvement in astronomy since Galileo's first spyglass in 1609. This potential rested first and foremost on its 94-inch (2.4 m) primary mirror, a precision-ground piece of glass so smooth that the largest bump on it was a mere millionth of an inch high. Hubble could see the stars on an American flag three thousand miles (4,428 km) away or a stretch limo on the Moon (provided that the limo was black). If our vision were equally sharp, we would be able to read license plates at twenty miles (32 km).[5] The telescope's ability to point precisely and stare steadily at distant targets was also key to unlocking its scientific potential, and Hubble's performance specifications on these points were impressive indeed. They were the equivalent of holding a laser beam steady on the head of a dime two hundred miles away for hours on end. It boggled my mind to think that a spacecraft zooming around the Earth at orbital velocities could hold that still. This remarkable instrument clearly had all the revolutionary potential Spitzer envisioned in 1946. My crewmates and I were so convinced of this that we joked about burning all of our college astronomy textbooks at our postlanding party, since Hubble was sure to render them obsolete in very short order.

Hubble is also an amazing piece of engineering. It is the size of a school bus, weighs 12.3 tons (11,100 kg), and contains some 400,000 parts and 26,000 miles (42,000 km) of wiring.[6] Those statistics alone make it a standout among scientific satellites, but what really sets Hubble apart—and makes it radically different from anything built before or since—is that it was designed to be maintainable. Not in a specialized satellite maintenance facility on Earth, but by two people in bulky spacesuits while orbiting hundreds of miles above the Earth. That's like working on your car while wearing an inflatable sumo wrestler suit and boxing gloves, with the added twist that your tools float away if you let go of them. Where did such a wild idea come from? How did a bunch of people who had never been in orbit or worn a spacesuit figure out what "maintainable" meant in a microgravity environment, and then incorporate all that into the telescope's design?

Those are the questions that led to this book. They propelled me to dive deeply into the written record of Hubble's history and to reconnect with some of the people who lived the journey and did the work of designing the world's first maintainable spacecraft.[7] Tracing how the idea of maintainability arose and evolved into the complex machine I would soon see in California gave me a much richer understanding of how both the telescope and my five years of work on it fit into the larger saga of learning to live and work in space.

✷

Before I could trace the arc of this idea throughout Hubble's history, I had to make sure I understood just what maintainability is. What makes something maintainable—able to be sustained or restored to proper operating condition—was not something I had ever thought much about. Most of us only think about maintainability when the breakdown of something we rely on shoves it in our face and makes us notice its absence, as when we have to take apart half of an engine to get at a failed part. My own experience gave me an instructive example. I once made the mistake of thinking I could repair the radio antenna on my car. This was an old-style antenna, common back in the 1980s, that automatically telescoped out when the radio was switched on and retracted when it switched off. Mine had quit retracting fully. After a day and a half spent dismantling the trunk compartment, wrestling with the spool of wire that drove it up and down, and swearing at a nut that would not come loose, I gave up and went to the dealer. It took the professional mechanic all of five minutes to fix it, because he had specialized tools designed just for this problem and had mastered the maintenance techniques and tricks of the trade that were peculiar to this task. That antenna was maintainable in principle, but not by an amateur in a home garage. Little did I dream during that frustrating weekend that just a few years down the road I would be producing for a space telescope the same trio of things that made my auto mechanic so efficient: reliable tools, proven procedures, and the knowledge known as tricks of the trade.

That simple example gives us a window into what goes into making a system maintainable, be it a simple car antenna or a massively complex space telescope. There is clearly an architectural dimension to maintainability: decisions have been made about which components need to be repairable, where to place those components, and where the repair will be done. Other fundamental principles come next, the first of which is to ensure the safety of the maintainer. This is done by designing the system so that any failure leaves it in a state that is safe to work on and including appropriate labels and warnings. The remaining principles for maintainable design are simple ones that are familiar to anyone who has ever done the least bit of home or car repair: things are easier to repair (more maintainable) when the layout of parts is simple and uncluttered, no healthy units need to be disturbed to get at a failed unit, the fasteners and connectors are standardized, and few unique tools are needed. Time and cost come into the equation through decisions such as how much downtime is tolerable, what the frequency of planned maintenance will be, and the size of the spare parts inventory needed to sustain the chosen maintenance strategy. Understanding and quantifying the interaction between all these variables—criticality, probability of failure, service interval, ease of repair, operational impact, cost of repair, and so forth—is the work of maintainability engineers. Maintainability is not something that can be added on once a design is complete, like a coat of paint on the outside of a house. Since it rests on fundamental properties such as the reliability of the parts and components built into the system, it must be an integral factor in the design process from the very start.

When did the idea of maintainability arise, and why was it considered such a vital feature for Spitzer's orbiting observatory?

*

It took nearly two decades for Spitzer's idea of a large space telescope to gain significant traction. The space age got off to a rocky start in the United States. The entire country, shocked and alarmed by the launch of Sputnik, was then dismayed to watch our own nascent space effort suffer a long string of embarrassing rocket failures. The first American satellite, Explorer 1, finally got into orbit in January 1958.[8] Seeking to focus the country's efforts and accelerate progress, President Eisenhower proposed in April of that year the formation of a new civilian agency that would be dedicated to the research and development needed to conquer space. Congress responded by passing the National Aeronautics and Space Act in July. The agency that would eventually produce an orbiting observatory very much like the one Spitzer envisioned, the new National Aeronautics and Space Administration (NASA), began operations on October 1 of that year.

Astronomy was an early focus of the fledgling agency's scientific efforts. By 1963, just five years after NASA opened its doors, an Orbiting Observatories program was up and running at the Goddard Space Flight Center (Goddard) in Beltsville, Maryland, with three teams working to produce satellite observatories that would study the Sun, Earth, and stars. Many astronomers at the time were leery of space-based astronomy. Rockets were very exotic and highly unreliable machines back then. Optical astronomers, those who worked with the visible light that reaches Earth's surface, preferred the annoyance of cloudy skies blocking their view for a few nights to the risk of losing finely crafted instruments in a rocket explosion; they were content to stick to their mountaintops. Those whose research required measurements in the infrared or ultraviolet regions of the spectrum had to get above the Earth's atmosphere, however. They had no choice but to work with the risky rockets and learn how to do astronomy in orbit.

The first of these Goddard satellites, Orbiting Solar Observatory-1, had been in orbit just over a year when the National Academy of Sciences' Space Science Board met in Iowa in June 1962. NASA had asked the board to weigh in on the research goals and strategy the agency should pursue after the first series of observatory satellites was complete. The panel assessed the potential of the emerging space-based capabilities to tackle the top-priority scientific questions of the day. On the subject of a large orbiting telescope, they observed that it would take an enormous investment to build a telescope of the size envisioned by Spitzer and noted, rightly, that a rigorous and comprehensive evaluation of the telescope's scientific value would be needed to justify such an investment. The consensus of the subpanel was to form a special committee to conduct this evaluation. A minority felt that even that was a step too far. Their dissenting statement provides an interesting insight into the state of science and technology at that moment: "At a time when not a single image of a celestial body has been obtained in a satellite, it is premature to convene a group to study a space telescope larger than the 38-inch telescope of the OAO [Orbiting Astronomical Observatory]."[9] It was a perfectly sensible but not very courageous statement, and far short of what Spitzer had hoped for.

Things had changed considerably when the Space Science Board met again at Woods Hole, Massachusetts, in the summer of 1965. In the intervening years, both Marshall and Goddard had commissioned teams of NASA and industry engineers to study possible telescope designs. These early studies were conceptual efforts framed by a few very simple starting points: take the mirror diameter and optical resolution proposed by Spitzer, plus current technology and the capability of existing American rockets, and solve for a telescope. Their purpose was to assess whether a large orbiting telescope was feasible with the technology of the day and roughly what it would cost to build one. None probed maintainability in detail, but all estimated what would be needed to keep such a telescope operating. A good example is a 1965 study entitled "Man-Tended Orbiting Telescope."[10] This described a telescope operated by shirt-sleeved astronauts working in a pressurized compartment at the back of the spacecraft (figure 4.1). Changing film was one of the most critical tasks envisioned for astronauts at the time, since electronic detectors did not yet exist. Another illustration in the same report showed an Apollo-style capsule serving as the crew delivery vehicle, although none had yet been built.[11] A third showed the telescope docked to a "manned orbiting research

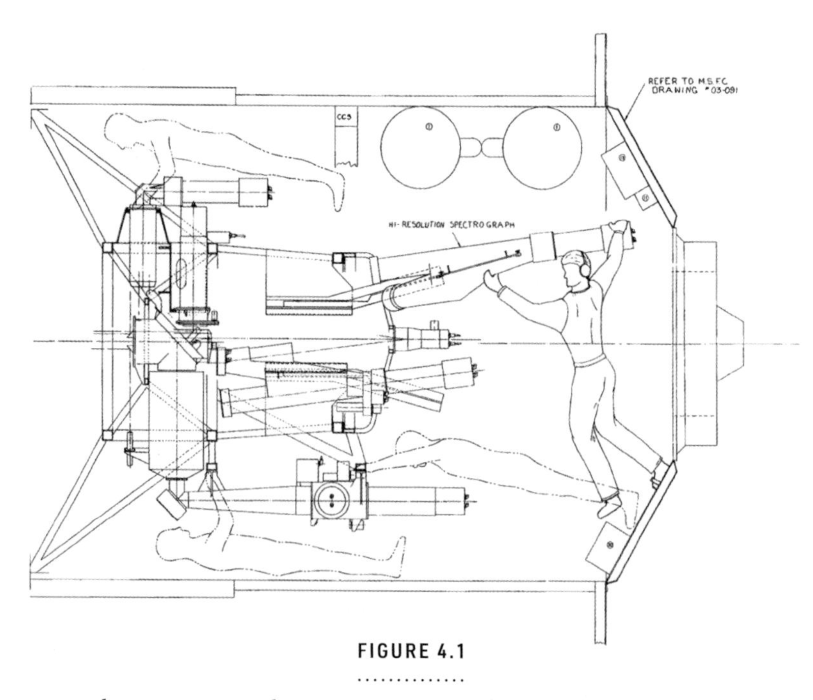

FIGURE 4.1

Early Large Space Telescope concept showing work locations inside a pressurized compartment. Source: National Air and Space Museum Archives.

laboratory," or space station. These were surely bold ideas to advance when the United States had only sixteen manned spaceflights and one fully successful spacewalk under its belt.

The question of a telescope such as Spitzer envisioned was again on the Space Science Board's agenda at Woods Hole. There had been no major achievements in optical space astronomy since the meeting in Iowa three years earlier, but the engineering studies noted above had demonstrated that putting a large telescope into orbit was feasible. Advances in stabilizing rockets with gyroscopes and in the techniques for producing space-quality astronomical instruments added further confidence. Against that backdrop, the board now concluded that "this large investment could provide a dramatic central focus for the optical space astronomy program, and that it would be an appropriate major space program for the Nation."[12] They went on to say that "a telescope of very large diameter ... and requiring the capability of man in space, is becoming technically feasible, and will be uniquely important to the solution of the central astronomical problems of our era."[13] This time, there was no dissent.

The space telescope had also garnered the support of a number of influential leaders in the young National Aeronautics and Space Administration by the time of the Woods Hole meeting, most notably NASA's chief astronomer, Dr. Nancy Grace Roman. The scientific potential of the Large Space Telescope, as it was then called, was very clear to Dr. Roman, and she was working hard to strengthen support for it both in the scientific community and on Capitol Hill. Other NASA leaders saw potential of a different kind. To Wernher von Braun at Marshall, such a telescope seemed another good use of the powerful and versatile Saturn rocket developed by his team. To Goddard's Dr. John Clark, it seemed the logical next project for the orbiting astronomical observatory team in his space sciences group, which had been building scientific satellites since before NASA was formed. Each man wanted to win the coveted "lead center" role on the Large Space Telescope, so his center would enjoy the scientific status and substantial funding that came with the title. This rivalry between Marshall and Goddard became a signature feature of the Hubble program. Generally constructive, sometimes troublesome but always fierce, it would influence the approach to maintainability that was designed into the telescope and complicate preparations for the first Hubble servicing mission.

By 1973, the telescope had enough basic engineering definition and political momentum for NASA to elevate it from the conceptual phase into the "design definition" stage, known as Phase B.[14] This is where general notions of how one might build a space telescope get turned into the specific design parameters needed to build a real spacecraft. The start of this phase coincided roughly with two other events that would have major impacts on the design of the telescope and the maintenance strategy. The first of these was NASA headquarters' decision that Marshall would be the lead center in charge of Hubble development and construction. Goddard was given the subordinate role of leading the development of the scientific instruments. This was surely a sensible division of duties and a politically savvy allocation of the funding that would eventually flow to the project, but it was also a decision that ensured that the two centers' intense rivalry would continue.

The second momentous event was President Nixon's approval in 1972 for NASA to proceed with the development of a new manned spacecraft called the space shuttle.[15] This new craft, still on the drawing boards, looked very good on paper. It promised frequent access to space—dozens of flights per year—at dramatically lower cost than other launchers. The combination of those factors and some key physical features of the shuttle itself seemed to open the door to a major change in the way space-bound satellites and scientific instruments had previously been designed.[16] The shuttle's roomy payload bay would remove the need for designers to squeeze the parts and components of a payload into the smallest possible volume, making preflight assembly, testing, and repair easier and cheaper. The projected high flight frequency and the presumed capabilities of its astronaut crews would make on-orbit maintenance and retrieval feasible. Early trade-off studies suggested that these factors could lower payload costs substantially, because designers could relax the extremely high reliability standards normally applied to space hardware and so avoid the cost penalty these imposed. One 1973 study estimated that the costs of the Large Space Telescope could be reduced by 30 to 40 percent over fifteen years.[17] More than a decade would pass before it became clear that the shuttle fell far short of these rosy cost and flight-rate predictions.

The shuttle and the telescope quickly came to depend on each other in both technical and political terms. The shuttle's on-orbit servicing capability shifted the notion of a large space telescope from a single-purpose instrument to a long-lived observatory that could carry out many different scientific campaigns. Mountaintop observatories had long operated in this fashion, but it was a big change for space astronomy. A long operating life would ensure a high scientific return on the huge investment needed to build the telescope. The ability to exchange instruments would allow the observatory to serve a larger segment of the astronomy community, keep pace with technology, and stay at the cutting edge of science. Maintainability would be key to all of this.

*

The task of figuring out how to meet the astronomers' ambitious performance demands and build the world's first maintainable spacecraft fell principally on the shoulders of lead systems engineer Jean Olivier.[18] Olivier was a Mississippi boy, the son of an oil-field truck driver, who had grown up working on cars and trucks. He was thrilled to get a slot in Chrysler's Automotive Institute after earning his mechanical engineering degree at Mississippi State University. Unfortunately, the country entered a recession just as he finished his master's degree at the institute, and Chrysler was unable to absorb the new graduates into their automotive business. Olivier went looking for open jobs in other parts of the company, intending to make his way back to his beloved cars when business picked up, and discovered that Chrysler had a contract with the Army's ballistic missile program in Huntsville, Alabama. It was not the work he really wanted, but the chance to move back to the South appealed to him. NASA came into existence shortly after he arrived in Huntsville, and Chrysler shifted him to designing launch facilities and ground support equipment for the new space agency's Cape Canaveral location. When NASA's newly formed Kennedy Space Center (Kennedy) took over those responsibilities in March 1962, Olivier decided to stay in Huntsville and found himself a job in Marshall's Advanced Systems Office, working under the legendary Wernher von Braun.

The objective of the design definition work that Olivier led was not to build anything, but rather to define the requirements and specifications that would control the building of this first-of-a-kind spacecraft. Some aspects of the telescope's basic geometry were already clear. The laws of optics dictated the size of the primary mirror needed to meet the scientific performance goals, and also how far apart the primary and the smaller secondary mirror had to be. The physical dimensions of the shuttle's cargo bay dictated the telescope's maximum length and diameter. The first thing Olivier and his team had to do was to come up with an architecture that satisfied these fundamentals, while also providing room for all the support equipment needed to power and operate the telescope—things such as batteries, computers, solar arrays, and communications antennas. Then, they had to specify the functions and attributes for each and every telescope system (e.g., guidance, electrical power, computing, data handling) and confirm that the ensemble would indeed produce a telescope that would perform as expected.

Three decisions made during the early design definition work established the basic scientific and maintainability features of today's Hubble Space Telescope. First, the mirror diameter was reduced from 3 to 2.4 meters (9.8 feet to 7.8 feet) so the telescope could fit into the shuttle's cargo bay. The second decision was of monumental importance for maintainability. Olivier's team chose what they called a "toroidal" architecture for the support equipment: a ring of equipment bays wrapped around the outside of the telescope, with equipment modules mounted inside. This approach allowed logical groupings of related components and ensured direct access to any module needing maintenance. Astronomer Bob O'Dell drove the third decision. O'Dell had left his post at the Yerkes Observatory in 1972 to join the Large Space Telescope team at Marshall as project scientist. He recognized that clustering the scientific instruments tightly together into a single "Scientific Instrument Package" (see figure 4.1 above), as Goddard then intended, would be disastrous for instrument maintenance and exchange. O'Dell forced the shift to the modular, self-aligning instrument architecture that has allowed Hubble's scientific capabilities to improve dramatically over the course of its life in orbit.

✷

In simple terms, any good telescope has to do three things: see sharply, point accurately, and hold still. The kicker was that this particular one, the Large Space Telescope, had to do all three exceedingly well while experiencing temperature swings from pizza-oven hot to deep-freeze cold and back again every ninety-seven minutes. That made Olivier's most wicked challenges the mirror, the pointing control system, and temperature management.[19] These would consume his team's attention and strain the project's budget repeatedly through the six years of Phase B. Olivier would use a combination of in-house and contract engineering teams to perform the quantitative analyses and tests needed to specify the functions and attributes of these three crucial elements.[20] As each bit of new information on one element came in, Olivier and his team would carefully examine how it affected all the others, making tough trade-off decisions or hammering out compromises where clashes arose.

Olivier's core team at Marshall did not devote much effort to maintainability during Phase B, and not just because they had their hands full with the mirror, pointing, and thermal problems. It was impossible to design detailed servicing schemes at this stage. As long as the telescope's configuration was still fluid and detailed design information lacking, the only work that could be done in this area would be exploratory and conceptual. Olivier left this to the three aerospace firms that had won Phase B design definition contracts and were likely to bid on building the telescope: Boeing Aircraft Company, Martin Marietta Corporation, and Lockheed Missiles and Space Company. The guidelines Olivier's team provided for these studies outlined NASA's high-level maintenance strategy at that time: no maintenance would be planned for the first two years of the telescope's life on orbit. Contingency maintenance during this early period would be limited to manual overrides of devices such as hinges and latches that were needed to either deploy the telescope or safely retrieve it and bring it back to Earth. The telescope would be returned to Earth for major refurbishment after the initial two-year period. During that refurbishment, it would be outfitted for "full orbital maintenance capability."[21] Thereafter, scientific instruments would be changed on orbit and the telescope would be returned to Earth for major refurbishment every five years.

There was another reason Olivier did not worry during Phase B about how to maintain the telescope: it was not his job to decide that. The "Request for Proposals" NASA would send out to potential bidders in 1977, when Congress had finally given the green light to begin building a real telescope, would lay out all the things that the contractor had to ensure the telescope could do, but would not tell them how to make it do them. It would provide the "What" without the "How," in other words. The sum and substance of its instructions on maintenance boiled down to "It shall be maintainable," plus the reminder that the high-level strategy was to combine "an on-orbit maintenance capability and earth return for refurbishment."[22] Hubble's maintainability would rest on the conceptual work done during Phase B by the successful bidder. This would turn out to be the Lockheed Missiles and Space Company in Sunnyvale, California.

✷

The Lockheed team tackling this challenge brought more than just predictive maintenance and reliability modeling skills to the task. Previous experience designing systems that could be maintained at remote military bases and in launch tubes aboard ballistic missile submarines made them highly alert to the difficulties that on-orbit maintenance would pose and especially to the human factors involved, such as cumbersome personal safety equipment, limited access and visibility, the need to tether every tool all the time, the value of error-proofing procedures, and the vital importance of ensuring the safety of the maintainers. Engineer Tom Fisher was one of the key people working to apply this experience to the development of the telescope's EVA maintenance features. Those features ranged from the fasteners used to attach components to the telescope and the connectors that carried electricity and data, to the handrails that astronauts would use to move around on it and the portable work platforms they would need in orbit. With no firm EVA data from NASA to guide them (the agency had only broad guidelines at that point), Fisher's group relied on basic human factors engineering principles, their own prior lessons learned, and common sense.

Two innovations provide good examples of the inventive and exacting work done during Phase B design definition. Both were driven by an interesting mix of the known and the unknown factors at that stage. The known factors, dictated by NASA, were that spacewalks would be no more than six hours long and involve just two astronauts. The unknowns were how much work would have to be done on a single spacewalk, which components would be targeted for on-orbit maintenance, and what the equipment failure rates would turn out to be. That combination of unknowns put a very high premium on simplifying tool interfaces for efficiency and reducing the overhead associated with getting set up for work.

Designer Henry Ford (no relation to the automotive family) decided to tackle the first of these by limiting the variety of fasteners used on the telescope. He set off in search of the smallest possible bolt that had the necessary high-tensile strength and a reasonable breakaway torque. In plain English, that means a bolt strong enough to withstand what he guessed the forces would be during a shuttle launch (the shuttle was still on the drawing boards, so there was no firm data), but not too hard to loosen with a wrench. Settling on a particular high-strength bolt with a double-height 7/16th-inch hexagonal head, Ford began his quest with the electronics installation team.[23] This team's job had always been to make sure that no electronics box ever came loose from its mountings; removing bolts in orbit had never occurred to them. After several rounds of argument and analysis, they agreed to use his chosen bolt on all of their units. Next, Ford went to the mechanisms team, responsible for things like the solar array and antenna hinges and latches, and finally the scientific instruments team. When he was done, he had collected the engineering data needed to prove the bolt was suitable for every application and had the commitment to use it from each system's lead engineer. With one more round of analysis and persuasion, he secured Marshall's blessing as well. NASA would later make this the standard EVA bolt on the shuttle and space station.

Tom Fisher's invention took aim at the time involved in worksite preparation. He knew the thirty-eight EVA worksites on the telescope better than anybody else, having produced by hand (there were no computer drawing or design apps back then) the precise and detailed engineering drawings of each location and the body positions that would allow suited astronauts to reach and operate all the maintenance fittings. The maintainability group used his drawings to develop full-scale worksite mockups and conduct in-house maintenance simulations, often hiring Tom Weaver, a spacesuit-qualified ex-NASA engineer, to stand in for an astronaut. They also booked time periodically on the telescope mockup in Marshall's neutral buoyancy simulator to confirm that the handrail and work platform locations they had come up with were truly workable in a weightless environment. Tom Styczynski was assigned the task of ensuring that the neutral buoyancy mockups accurately reflected the current design. When the Lockheed team felt they had a sound configuration, they would invite astronauts to evaluate it as well. As the tests progressed, Fisher saw the need for a portable EVA work platform, known as a foot restraint, that was more versatile than the simple one NASA envisioned using on the shuttle.

A portable foot restraint is to astronauts what gravity is to earthbound mechanics: the thing that allows them to anchor their feet so they can exert leverage or apply force to their tools. Try to turn a bolt in the microgravity of space without a foot restraint, and you will find that your free-floating body turns instead, because of friction between the shaft of the bolt and the surrounding metal. The shuttle foot restraint Fisher and company were using in their tests was basically a plate on a pole. It was designed to stick straight out from the structure to which it was attached. The foot plate could only tilt in one axis and could not be adjusted while in use. The neutral buoyancy tests confirmed that astronauts could not reach all the telescope worksites with such a rudimentary foot restraint, and that far too much time would be lost getting in and out of it to change its position. Fisher came up with the idea of adding pedals to the foot plate so astronauts could swivel their bodies with the tap of a boot. We would refine Fisher's initial concept through a succession of neutral buoyancy tests starting in 1985, eventually producing a highly versatile device that flew on every Hubble mission and is now in use aboard the International Space Station.

The maintenance concept that the Lockheed team developed combined rigorous reliability analysis, predictive maintenance modeling, and basic principles of human factors engineering. They had thoroughly assessed the criticality, reliability, location, accessibility, connections, interfaces, size, and weight of every component on the telescope, and identified the design principles and hardware standards that needed to be applied to ensure that the telescope would be maintainable both on the ground and in orbit.[24] The next challenge was to build it. That work was just starting when my TFNG classmates and I arrived at NASA.

5

MISSION PREP

Bruce McCandless and I had not crossed paths much since our Tinker-Toy EVA test in the water tank some five years earlier, but now we were sharing an office and joined at the hip on everything Hubble. My new crewmate was a fascinating character. The son and grandson of Medal of Honor recipients, Bruce graduated second in his class at Annapolis and went on to earn his Navy wings. After several tours as an aviator, including one aboard the USS *Enterprise* during the Cuban missile crisis, he went to Stanford for a degree in electrical engineering. NASA selected him as an astronaut in 1966 and then, just as it had done with us, put him to work supporting preparations for the Apollo missions that were soon to begin. He was the communicator in mission control (known as Capcom, for Capsule Communicator) for Neil Armstrong and Buzz Aldrin's moonwalk on Apollo 11 and became the world's first self-propelled human satellite when he flew untethered out of the shuttle's cargo bay in February 1984.

Bruce was a brilliant engineer, with a photographic memory and extraordinary capacity for detail. Brainstorming sessions with Bruce commonly involved him rattling off exactly which type and size of fastener or part the new gadget called for, what metals or synthetic materials were best suited to the purpose (along with what physical properties made them so), and which vendors, by name, would be good sources for each component. He had co-designed the shuttle Manned Maneuvering Unit and, in the years yet ahead of us, would design and patent several pieces of EVA equipment that astronauts continue to use to this day. One of the best examples of his inventiveness is a small device nicknamed for him: the McTether.[1]

"Everything shall be tethered at all times" is a cardinal rule of EVA. Application of this rule begins with the spacewalkers themselves. Each shuttle spacewalker carried two "safety tethers," one attached to each of the two metal rings near the hips of her spacesuit. One end of one of these would always be hooked onto a solid structure, such as the space shuttle or the work platform on the end of the robotic arm. Like a mountain climber, whenever she needed to change locations, she would tether to a new safety point before unhooking from the old one.

The tether rule applies to tools as well. Each and every item—the wrench handle, each individual socket extension, every other gadget needed at a worksite—must be tethered before it is moved. Each astronaut carried two shorter "wrist tethers" for such tool handling, one attached to a loop on the cuff of each glove. These were fine for moving one or two items, but not for handling the huge number of tools required for Hubble maintenance. It was clear to Bruce and me that far too much time and hand strength would be consumed just by the simple task of transferring tools from the tool box to the portable work caddies during Hubble maintenance spacewalks.

The reason for this was the "drop-proof" hook used on space shuttle tethers at the time (figure 5.1). A double-sided push button controlled a mechanism that prevented the hook from opening accidentally. Squeezing these buttons together between your thumb and forefinger released the lock, allowing the jaw of the hook to move when your other three fingers squeezed the handle. Anyone

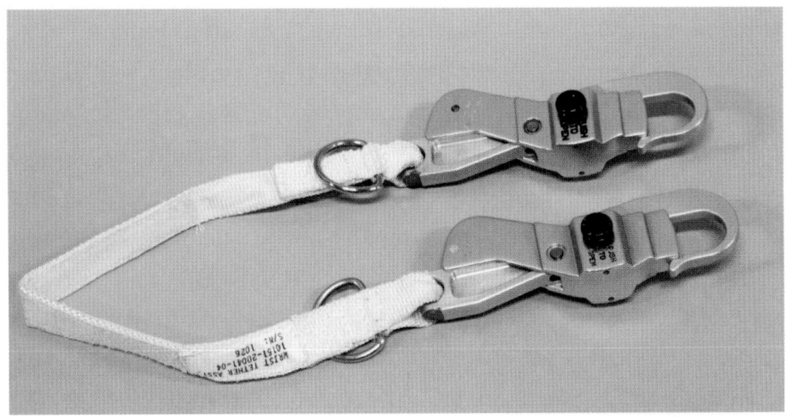

FIGURE 5.1

Space shuttle tether hooks (top) and T-shaped pip pin (bottom). The black circles in the center of the shuttle hooks are the buttons that unlock the tether. Source: NASA; Vlier Products.

who has ever touched a musical instrument knows how hard it can be to get three fingers working independently of the dominant thumb and forefinger. It is many times harder when wearing stiff and bulky spacesuit gloves that sap grip strength and remove any sense of touch. The tether hook had to be held just so to keep the buttons depressed while simultaneously squeezing the handle to open the jaw. Two hands were often needed to get the hook into the proper position, and the simultaneous pinch and squeeze actions taxed every hand and wrist muscle. This would just not do.

Bruce saw a simple solution to the tether problem. A T-shaped "pip pin," like the ones we used to align or connect pieces of equipment together aboard the shuttle, could be turned into a tool tether. The T-shaped pin would be much easier to grasp and operate, needing only three fingers and a simple, syringe-like motion. Tethering a tool would be like spearing it with the T-shaped pin. Two things were needed to turn this idea into reality: creating a matching socket small enough to add to each tool, and persuading the Hubble program managers spend the time and money to put one on each item. The first one was already done. Michael Withey, a contractor working with the crew systems group at Johnson and every bit as brilliant a designer as Bruce, had produced a version that Bruce tested on his first flight and that was used on the Solar Max mission.[2] Michael and Bruce quickly adapted the first socket design to fit on all of the Hubble tools. Armed with this and the task timeline data from our neutral buoyancy tests, we set about convincing the Hubble project managers to add McTether pins and sockets to every tool and work caddy in the maintenance kit.

*

The Lockheed team assembling the telescope in Sunnyvale scheduled a full day in mid-April 1985 for a series of familiarization briefings, hardware inspections, and publicity photo sessions with us, the newly named Hubble deployment mission astronauts. The briefings would give us an overview of the telescope's design history and scientific objectives, but focus primarily on the interactions we were meant to have with it on our mission. These interactions would not include a spacewalk, if things went according to plan, but both the presentations and the tour that followed would cover the EVA accommodations nonetheless. After three months poring over design reviews and test reports, I was eager to see the telescope and meet the team that was getting it ready for launch.

In 1985, Lockheed's Sunnyvale plant occupied a four-hundred-acre parcel of land nestled between the southern edge of San Francisco Bay, a major highway and the runways of what was then Naval Air Station Moffett Field.[3] Some twenty thousand people worked there at the time, most of them on classified Air Force or Navy space programs. Spacecraft assembly and testing took place in an industrial area on the western edge of the site, abutting the Navy airfield and Air Force Station Sunnyvale, which was famous for its large windowless building known as the Blue Cube (figure 5.2).[4] Most of the Lockheed campus consisted of nondescript low-rise office park buildings, laid out on streets with unlikely names such as Caspian, Caribbean, Baltic, and Gibraltar. My father had worked on Navy missile programs in one of those featureless buildings since moving to the area in 1973. The Lockheed component of Team Hubble, the group that had designed and built the amazing machine we would deliver to orbit, was waiting for us in Building 579 on Orleans Drive.

Program manager Bert Bulkin kicked off the proceedings. Bert's accent announced immediately that he was a Brooklyn boy, although he had been in California since the late 1940s. He had worked at Lockheed's famed and super-secret Skunk Works during summer breaks from his studies at UCLA and joined the group

FIGURE 5.2

The "Blue Cube" at Air Force Station Sunnyvale. Source: Wikimedia Commons.

as a draftsman after getting his aeronautical engineering degree in 1951.[5] By the late 1950s, he had risen to design engineer and left Lockheed to work on the country's first reconnaissance satellite, code named CORONA. Bert returned to Lockheed in 1965 to work on another early spy satellite called HEXAGON and had been a key figure in the company's Hubble work since the start of the Phase B design definition work in 1972. Now he had the tough job of making everything come together to deliver the high-performance, maintainable spacecraft NASA had ordered, without any more of the schedule slips and budget overruns that had plagued the program in prior years.[6] Bruce and I would have fairly little direct contact with him in the years ahead, but every issue or decision about the work that needed to be done before launch to ensure Hubble was maintainable on orbit would cross his desk.

The most physically striking man in the room was Frank Costa, the czar of Hubble's electrical system. That was surely not his official title, but it was how everyone referred to him and, as I would quickly learn, very much how he saw himself. "Mussolini" and "Icepick," nicknames his colleagues had given him, aptly captured both his gruffly imperial manner and Italian heritage.[7] His imposing stature, booming voice, rugged hands, and thick mane of dark hair were all notable features, but he was most famous for his eyebrows. Luxuriant is the only word that does justice to the wide, thick, and amusingly unruly patches of black hair that crowned his eyes. They were so extravagant that he had to cover them with a custom-made piece of protective gear whenever he went into the super-clean chamber where Hubble was being assembled. Needless to say, he was most widely known as "the guy with the crazy eyebrows."

Frank was a wiring genius. He got his start working for a Hollywood recording studio while taking electronics classes at a junior college in Los Angeles. Then, armed with his associate's degree, he got a job wiring submarine launch control consoles for the Firestone Company's Guided Missile Division. He signed on with Lockheed's then newly formed Missiles and Space Division and moved north to Sunnyvale in 1960. It is no stretch to say that Frank Costa personally laid every inch of wire in the Hubble Space Telescope. In a 1989 interview with historian Joseph N. Tatarewicz, he describes how he studied the bare structure of the telescope for about a week, mentally threading the cable along "corridors that I had in my mind," before he began to produce the layout drawings.[8]

We would be under Frank Costa's very watchful eye any time we went near an electronics module with tools in hand.

The Toms—Tom Dougherty and Tom Styczynski—were the two most important people in the room for Bruce and me. Dougherty was the manager in charge of Hubble M&R, responsible for everything associated with servicing, from prelaunch preparations to the logistics program that would be needed to sustain the observatory throughout its fifteen-year lifetime. He had joined Lockheed immediately after finishing his degree in electrical engineering at City College of New York in 1961. A native of New York City who had never needed a car while growing up, Dougherty essentially learned how to drive on the trip west to Sunnyvale. The word among his M&R team colleagues was that he had not learned terribly well, as proven by the fact that his wife always drove the family's recreational vehicle. He reminded me very much of my New Yorker father, with similar facial structure, hairstyle, accent, and easy laugh.

Tom Styczynski was then helping Dougherty put together Lockheed's proposal to provide engineering support for the servicing missions that would occur throughout the telescope's lifetime (the company's contract at the time extended only through the initial on-orbit checkout period). Styczynski was a mechanical engineer and sports car aficionado whose first job upon joining the company in 1974 had been designing heavy-duty tactical military trucks and armored personnel vehicles—a dream job for someone who had wanted to be an automotive engineer since age twelve.[9] His engineering interest had been spawned by childhood trips to Chicago's train yards with his father, a day laborer on the Baltimore & Ohio Railroad. The flows of animals and rail cars in the stockyards and the locomotives in various stages of disassembly in the roundhouse made him wonder how such things came together. Cars later replaced trains as his primary interest, thanks to the *Hot Rod* and *Motor Trend* magazines his sister and brother-in-law gave him as a youngster. His assignment to the Hubble project in 1978 came about through a chance parking lot encounter with his supervisor, Henry Ford. At the time, Lockheed's Phase B design team was preparing the proposal the company would submit to NASA as its bid to win the Hubble construction contract. Ford, who had been helping Tom Fisher's group with the human factors and crew systems work, thought Styczynski's experience as a car mechanic would be helpful on the servicing aspects of the design. Tom S. joined the Hubble team immediately after the telescope construction contract was awarded in October 1978.

*

We had met the important players on the Lockheed team. All the briefings were done. It was finally time to meet Hubble.

Getting into the chamber where the telescope was being assembled involved a process that was almost as intricate as suiting up for a spacewalk. Hubble was housed in an enormous, one-of-a-kind facility called the Vehicle Assembly and Test Area, or VATA for short. The VATA was built so that very large satellites could be assembled and tested in extremely clean conditions. It was as tall and wide as a basketball court—ninety feet by fifty-five feet—and one hundred twenty feet long; large enough to hold 437 shipping containers. It had a massive overhead crane that could raise a twenty-ton load to a height of almost seventy feet, and a sophisticated set of movable work platforms that could be adjusted to provide access to every part of a spacecraft.

The cleanliness inside the chamber was absolutely critical for Hubble. An unimaginably small amount of contamination or dust on the mirrors would reduce the telescope's optical performance significantly, potentially rendering the resolving power of this very expensive space instrument no better than a much cheaper ground-based telescope. In technical terms, the VATA was a Class 10,000 clean room. In plain English, that means that there was 1/10,000 as much dust in the air surrounding Hubble as there would be in normal household air. The building's air handling system kept it that way. The pressure inside the chamber was slightly higher than the pressure outside, so that dirty outdoor air could not leak into it. All of the air was drawn into the chamber through one of the ninety-foot by fifty-five-foot walls, which was covered with an array of huge fans and high-efficiency particulate air (HEPA) filters. The fans created a smooth airflow along the length of the chamber that was just fast enough to keep aloft the few minute particles that might have escaped the filters. All the air in the immense chamber was refreshed every ninety seconds, about half the time it takes to refresh the air inside a passenger jet. As a final measure, Hubble was assembled at the upstream end of the chamber, closest to the filter wall. Every person, tool, or piece of equipment needed to work on the telescope came in downstream of it and stayed downstream as much as possible to reduce the contamination risk. People going in and out of the VATA were the major threat to the chamber's extreme cleanliness, so the rules and procedures for entry were very detailed and very strict. We had been warned beforehand that our preparations for visiting Hubble had to begin the moment we woke up in the morning. The use of skin lotion, after shave, cologne, makeup, nail polish, and lip balm was strictly banned on VATA days.

We entered the two-story building that housed support functions for the test chamber and followed our guides toward the gowning room. We had to clear a number of hurdles just to get inside that. The clean room clerk, posted at a window just outside the door, grilled each of us in turn about the banned substances and inspected us for jewelry we would have to take off or, in the case of wedding rings, put tape over before gowning up; no unnecessary hard objects were allowed near the telescope. Only when he was satisfied would he allow us to sign the entry log. With that done, he handed us a stack of vacuum-sealed plastic packages containing our clean room garments. These consisted of a hood that covered head and neck like a medieval knight's chain mail hood, a full-body coverall, and knee-high boots to go over our shoes. Men with facial hair got an extra mask that snapped inside the hood to cover their beards or mustaches, and Frank Costa got his special eyebrow shield. The clerk then watched like a hawk to make sure we each ran both of our shoes through the powered shoe brush and then stomped both feet on the sticky door mat before stepping into the gowning room.

Another intricate ritual began once we were inside. The gowning room was a rectangular area divided into two zones by a wide, L-shaped bench. The side of the bench we entered on was the dirty zone. The top of the bench and everything to the other side was the clean zone. Our first task was to put the clean room garments on over our clothes without letting anything clean touch something dirty, or vice versa. If a clean garment so much as brushed the dirty-side floor, it had to go into the laundry bin and we had to go back outside to get a fresh one. After some comically awkward gymnastics, everyone in our group was finally standing fully gowned on the clean side of the room. The next step was to wrap an electrical grounding strap onto one wrist and clip it briefly onto a metal grounding point, so that any static electricity that had built up in our clothing would be discharged. The telescope's sensitive electronics could be badly damaged by even the small doses of static we experience in everyday life, so everyone working on the telescope had to make sure they never ever touched Hubble without first attaching their grounding strap to a nearby piece of metal. Next we donned latex gloves and taped them to our sleeves and taped the top of our boots to the legs of our coverall, all to ensure that none of the dust and fibers trapped in our clothing would escape while we moved around. Our guide inspected us to make sure we had everything right and led us toward the final hurdle standing between us and the VATA chamber. This was a small, airlock-like chamber that is aptly called the air shower. One by one, Bruce and I stepped onto another adhesive mat to remove any bits of bad stuff that might have somehow gotten onto the bottom of our just-unwrapped clean room booties and stepped inside, followed by our guide. A dozen nozzles, resembling the air jets in a hot tub, lined the chamber's walls and ceiling. These came to life with a roar as soon as the door to the gowning room closed. Following our guide's example, we moved about just as we had during our morning showers at the hotel, turning around and lifting our arms so that the flowing air hit all parts of our body and blew off any bits of dust or fragments of fiber that might have been clinging to our bunny suits.

A loud click at the end of the air blast announced that we were finally fit to enter into the presence of the Hubble Space Telescope. Our guide opened the door, and I stepped into the vast clean room chamber. We think nothing of looking up at a nine-story building in our age of skyscrapers, but I felt like an ant when I walked into the enormous, nine-story-high vault of the VATA (figure 5.3). It seemed brighter than daylight in there. The floor was as smooth as glass and more than clean enough to eat from. Every surface of the chamber gleamed brightly. Nothing outshone the telescope, however. There it was, some fifty feet away from me, a stunningly shiny silver creation standing upright amid the white access platforms. None of the artists' renderings and engineering drawings I had pored over during the preceding months prepared me for the beauty of the sight. Hubble looked like a piece of precious sterling silver. Put it in a light blue box and it could be a gift from Tiffany's.

FIGURE 5.3

The Hubble Space Telescope in the VATA clean room. Note the people at lower left for scale. Source: NASA.

I knew the telescope's dimensions by heart (fifty feet long, just shy of fifteen feet in diameter at the bottom, and thirteen feet in diameter at the top), and was familiar with its comparison to a school bus, but I was still struck by how large it was. The two grapple fixtures that we would use to deploy the telescope, and that future crews would use to capture it again for maintenance, looked tiny on this huge spacecraft. They were dwarfed by the four-foot-diameter dishes on the high-gain antennas that would beam the astronomical data back to the ground. As we moved closer to Hubble, the basic maintainability features incorporated in the design shifted from the black-and-white of our engineering documents to full color. The bright yellow handrails that marked the EVA pathways and worksites for on-orbit servicing stood out most vividly. I imagined what it would be like to pull myself along them while flying hundreds of miles above the Earth and wondered if I would ever get the chance to find out.

Bert Bulkin and Tom Dougherty were no doubt leading our group, with Tom Styczynski and others in tow, but I honestly remember nothing of that day in the VATA but Hubble. We climbed to the highest level of the work platforms to begin our top-to-bottom tour of the spacecraft and its maintainability features. The yellow EVA handrails that stood out from afar may have been the most visible of these elements, but they were far from the only important ones. My first thought was that the telescope's intrinsic maintainability features were just commonsense approaches to simplifying both ground assembly and testing and on-orbit maintenance: install key equipment modules inside easy-to-open compartments; use a standard fastener for all equipment mounts; leave enough room for technicians and spacesuited astronauts to get their hands onto connectors and other fittings; ensure that removal of one component doesn't require removal of any others. These are basic principles taught in engineering classes and familiar to anybody who has ever worked on cars or done home repairs, but I marveled at the foresight involved. When the idea of having astronauts maintain an orbiting telescope arose, manned spaceflight was in its infancy. Just two short, simple spacewalks had been done, one each by the Soviet Union and the United States, and the former had nearly ended in disaster.[10] Yet, in the two decades since, a team of NASA and corporate engineers who had never flown in space or worn a spacesuit had produced a vehicle that was remarkably well-suited to on-orbit maintenance.

*

Let's digress for a moment to recreate the tour we had that day, using the artist's concept in figure 5.4. The flat surface at the top of the telescope is Hubble's lens cap, technically known as the aperture door. It is hinged along the flat side and held shut by a latch at the top of the curved side (just above the first "A" in the NASA logo). Manually releasing the latch and cranking the door open were two of the backup or contingency EVA tasks that Bruce and I might be called on to perform during our mission. Below this, at about the same level as the NASA logo, two yellow structures on struts stick out from either side of the telescope. These milk stools, as we called them, are bumpers designed to keep the thirteen-foot-diameter section of the telescope's barrel (technically, the light shield) from hitting the sides of the shuttle's cargo bay during launch.

The twin solar arrays come next, one on each side of the barrel immediately below the milk stools. The illustration shows them in launch configuration, with the forty-foot-long golden blankets furled like rolled-up window curtains, and the masts latched onto the side of the telescope. Releasing the two latches that hold the masts in place, pivoting the masts down ninety degrees, and unfurling the blankets were more tasks that might fall to Bruce and me. The two high-gain communications antennas are also mounted to this segment of the telescope's barrel. One is shown in the deployed or ninety-degree position on the side with the NASA logo; its twin on the opposite side is left out for clarity's sake. As with

FIGURE 5.4

Artist's rendering of Hubble's modular architecture and key On-Orbit Replaceable Units (ORUs). Counterclockwise from upper left, the modules shown outside the telescope are the Science Instruments Control and Data Handling unit; six battery modules; an axial science instrument; the Wide Field/Planetary Camera (WFPC); a second axial science instrument; one of the three Fine Guidance Sensors; and two gyroscopes. Source: NASA.

the solar arrays, the antenna latches and hinges were designed to allow for manual override if the motorized systems failed.

Hubble's diameter changes from thirteen to fifteen feet just below the solar array and antenna hinge mechanisms to accommodate the scientific instruments and the electronic equipment needed to power and operate the telescope. The various system modules are mounted in two tiers of compartments, or equipment bays, that wrap around the telescope (Jean Olivier's toroidal architecture). The exterior access doors are open or missing in figure 5.4 in order to show the arrangement of equipment within the bays and on the inside faces of the doors. Three guidance sensors and five scientific instruments occupy the telescope's aft-most section, the two lowest segments in the figure. The four refrigerator-sized scientific instruments that occupy the lowest portion of this section are known as the axial scientific instruments, because the light beams entering them are parallel to the long axis of the telescope. The thin segment between the axial instrument compartment and the electronic equipment bays is occupied by three Fine Guidance Sensors and Hubble's fifth scientific instrument, the Wide Field/Planetary Camera, which, along with its successors, is responsible for most of the dazzling images that have propelled the telescope into pop culture.[11]

✷

Back in Houston, Bruce and I caucused with the EVA specialists who were assigned to support the deployment mission and would become the Johnson Space Center element of the Hubble M&R team. At the time, the EVA section of the shuttle mission control team was a small, loosely organized bunch that worked EVA operations end to end. Instead of having separate groups working tool production, procedure development, crew training, and, finally, in-flight operations, we had a quartet of engineers—Kitty Havens, Jim Thornton, Robert Trevino, and later Sue Rainwater—who worked it all as a seamless team. One or more of them would monitor every neutral buoyancy simulation as if it were a real EVA and join us in the VATA whenever we tested tools on Hubble. Between those touch points with us, they produced the documents and engineering drawings needed to ensure that the feedback from these events reached the right people and resulted in the necessary changes. Who better to staff the EVA console in mission control during our flight than the engineers who helped build every tool, had written every checklist, and were intimately familiar with the skills and perspectives of the mission's two spacewalkers?

We faced a three-pronged challenge. First, we had to make absolutely sure everything was one hundred percent ready for the deployment mission. Meeting this bare minimum target would require producing and testing the relatively small complement of tools needed to handle the latch, hinge, or motor failures that could threaten the successful start of the Hubble mission. Second, we had to do everything we could while Hubble was still on Earth to ensure that no future Hubble maintenance crew ever found themselves on a spacewalk with equipment that did not fit or work as needed. Tool failures like the ones that happened on the satellite repair and retrieval missions in 1984 would be unforgivable on a spacecraft that had been designed from the outset to be maintainable. The only sure way to prevent this was to produce a complete set of all the tools needed for the full suite of maintenance tasks and test each one on every Hubble bolt or connector. That would also let us confirm that a suited astronaut really could reach all the bolts and connectors on the real telescope, something it was impossible to do with the rather crude mockups we had in the neutral buoyancy simulator.

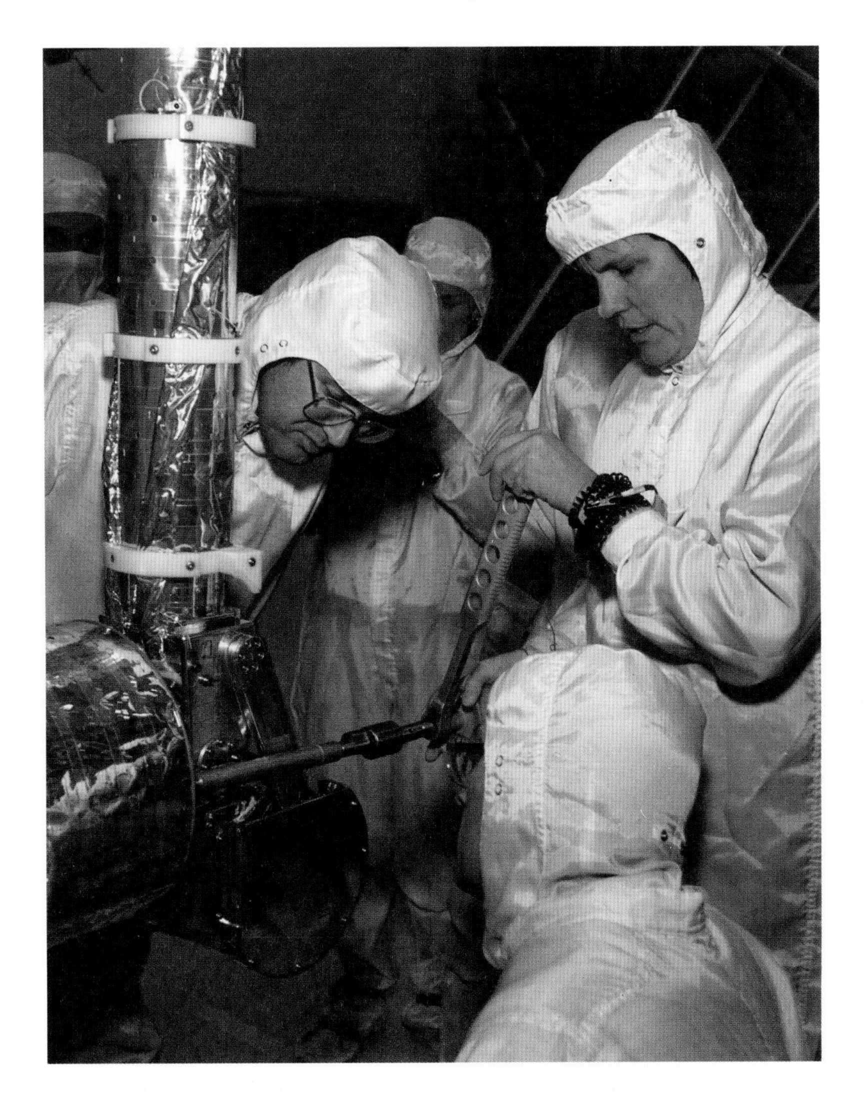

FIGURE 5.5

Kathryn Sullivan (right) checks the fit of an EVA wrench on one of the solar array manual override mechanisms during a test in the VATA, as Bruce McCandless (bottom) and British Aerospace Company engineer Barry Henson (top) look on. Source: NASA.

The third prong of our challenge resulted from a major change in maintenance strategy adopted in the 1983–1984 timeframe. The strategy until then had been to bring the telescope back to Earth every five years for major refurbishment. On-orbit maintenance would be limited to swapping scientific instruments and replacing the small number of vital components that were known to have either limited lifetimes (like solar arrays and batteries) or inherently low reliability (such as gyroscopes and tape recorders). The folly of this strategy became clear after just a few years of shuttle operations. High costs, unreliable launch schedules, and contamination concerns combined to convince both the scientific community and NASA that Hubble would end up in a museum rather than back in space, if it were ever returned to Earth. As a result, ground return was officially dropped as an option; all maintenance would be done on orbit by space-suited astronauts. This shift sent the reliability engineers scrambling to determine which electronic modules were most critical to modify for on-orbit replacement, while the maintainability engineers began thinking about how to convert as many of them as possible into ORUs—On-Orbit Replaceable Units—or at least make them more "EVA friendly."

Bruce and I were behind this effort one hundred percent. We were certain that NASA would send a crew to try to fix any component that threatened to end Hubble's scientific mission, regardless of whether it was EVA compatible or not. We also believed that if the fix did not work, the headlines would read "NASA Astronauts Fail to Fix Hubble," not "Late Strategy Shift Sealed Hubble's Fate." For the sake of those future maintenance crews, we had to push as hard as possible for changes that would increase the odds of success. Time was not on our side. Lockheed was scheduled to start packing up Hubble for shipment to Florida about three months before our October launch date. That left us barely twelve months to create the tools, equipment, and procedures that would make Hubble fully maintainable on orbit.

*

The combined Lockheed and Johnson M&R teams laid out an ambitious schedule for the remainder of 1985 and first quarter of 1986, driven by tool-production schedules, the completion of improvements to the mockups in the neutral buoyancy water tank at Marshall, and events in the workflow at Sunnyvale that would allow us to test our equipment on Hubble. To my great delight, the plan included two week-long rounds of development runs in the water tank, one in October and a second in November. I could not wait to get back in a spacesuit and do some zero-G work, even if only in an underwater simulation.

By the time we convened at Marshall for the November neutral buoyancy simulations, Tom Dougherty had shuffled the Lockheed M&R team. Realizing that he would need a deeper talent bench to handle the simultaneous demands of prelaunch M&R work and long-term maintenance preparations, he shifted Styczynski onto the long-term task and assembled a new team for the prelaunch work. The talent for new group this came from the company's logistics division, then headed by Lee Dove. Dove sent him a retired Army helicopter pilot who was embarking on a second career and two young engineers fresh out of San Jose State College's aeronautical engineering program. The upcoming work at Marshall would be this trio's first meeting with Bruce, our Johnson EVA team, and me, and their first exposure to zero-G simulations. Little did any of us dream that they would become the mainstays of Hubble servicing for the ensuing twenty-four years.

The retired Army pilot and ringleader of this new group was Ron Sheffield (figure 5.6). A tall, lanky, fiercely loyal Sooner,

FIGURE 5.6

Ron Sheffield as a Lieutenant Colonel in the 82nd Airborne Division. Source: Sheffield personal collection.

Sheffield graduated from the University of Oklahoma with a degree in mathematics, an ROTC commission, and Linda, the love of his life. His twenty-year Army career included three combat tours, during which he earned a Silver Star, a Bronze Star, a Distinguished Flying Cross, fifty-nine Air Medals, and a number awards for gallantry and valor from the government of Vietnam. He had been hired into Lee Dove's maintainability group scarcely a week before Dougherty came looking for people to beef up the Hubble M&R team. The steel that Sheffield's Army experience had forged in him was paired with an easygoing temperament and a playful sense of humor. This struck me as an ideal combination for the man whose task as Hubble's new M&R manager was to upgrade every electrical and mechanical component on the telescope so that it could be replaced in orbit. It certainly made him a good leader for the young engineers who went over to the Hubble program with him.

Peter Leung and Brian Woodworth were the brand-new San Jose State graduates in the group. Brian was a local Bay Area boy whose ambition in life was to become an engineer, build his dream house in the Almaden Hills south of San Jose, and raise a family. An entry-level job at a bustling company like Lockheed was a great first step in that direction. Peter, on the other hand, had a much more exotic past. Born in mainland China, his family managed to escape to Hong Kong when he was about two years old. Peter's father, a farmer who had been imprisoned in a labor camp after the Communist Party took over, could no longer work, leaving the family to eke out an existence on what money his mother could earn. Peter's early education was paid for with funds sent to his parents by an older brother who had immigrated to the United States. When that brother became a citizen, he petitioned to bring his youngest brother to join him in the US for schooling. Peter arrived with modest English language skills but a fierce passion to build a good future for himself. Giving freshly minted graduates like Brian and Peter big roles on major projects like Hubble was a deliberate strategy in the maintainability group at Lockheed, intended to infuse fresh perspectives into long-standing teams and jump-start the careers of promising young engineers. It certainly worked that way with these two. They quickly became so integral to our work that Bruce and I thought of them as Sheffield's right and left hands, with Brian focused on tool design and testing and Peter on ensuring our training hardware mirrored the telescope faithfully.

The tests scheduled for October and November had two objectives. The first was to start the iterative process of turning the engineers' preliminary, generic notions about how to perform the deployment mission maintenance tasks into detailed EVA procedures tailored to Bruce and me, reflecting our height, our arm length, and the way we worked in the spacesuit. The second was to determine the specific portable foot restraint setups that would be needed to work at every one of the telescope's EVA worksites. Tom Styczynski and the Lockheed team, with periodic advice from Bruce and other visiting astronauts, had identified the worksite locations during the design phase and installed sockets for portable foot restraints at each one during assembly. It was vital that we confirm that these provided adequate reach and access while the telescope was still on the ground, in case changes needed to be made. At first blush, this sounded like one of the necessary but rather mundane chores that goes into the development of every spacewalk. Little did I dream as I began my first run in the water tank that it would turn into a classic example of how vital hands-on experience and iterative attempts are to the inventive process, and would result in a device—an adjustable portable foot restraint—that would become standard on-orbit operating equipment for decades to come.

Three strands of experience fed into the months-long creative process that would begin with these neutral buoyancy runs. One was many years of experience working with the shuttle's standard portable foot restraint. This simple device consisted of a foot plate, with toe loops and heel clips to grab a spacewalker's boots, mounted on a straight metal rod. The rod ended in a large hexagonal pin that fit into the matching sockets installed around the shuttle's cargo bay. The six faces on the hexagonal pin, plus two joints located just below the foot plate, gave astronauts some ability to fine-tune their working position in three axes: pitch (head forward or back), roll (tilting toward left or right shoulder), or yaw (swiveling from side to side). Time was the greater problem with this design. Even with each astronaut's position settings known in advance, it took a long time to set up the foot restraint. Worse, you had to get out of it and maneuver to the underside to make even the slightest adjustment, then maneuver back to get into it again. Kitty Havens and Robert Trevino, then the Hubble EVA leads from Johnson, had worked with this device since the start of the shuttle program and knew we could not afford to spend so much time on foot restraint adjustments during Hubble maintenance spacewalks.

The second strand was the work that Lockheed had done during Phase B to maximize EVA time efficiency. Tom Fisher would have been familiar, at least in broad outline, with the foot restraints used in the earliest NASA spacewalks and the deficiencies of the shuttle unit. He, too, recognized how precious EVA time would be and so had conceived of a foot restraint that astronauts could swivel without getting out of it. The prototype of Fisher's design would get its first practical evaluation during these late 1985 neutral buoyancy tests.

The third and final strand was hands-on experience with the real Hubble Space Telescope—the "flight article," as we called it. For the first time, on-orbit maintenance operations would be evaluated in simulated zero G by veteran astronauts, EVA experts, and maintainability specialists who were intimately familiar with details that neither engineering drawings nor crude underwater mockups can convey. We knew, for example, just how far into or across equipment bays astronauts would have to reach to perform a task and what it was like to work with a bunch of tools attached to the bracket on the front of the spacesuit. These operational insights gave us a much keener understanding of the flexibility and efficiency Hubble spacewalks would require and how valuable an easily adjustable foot restraint would be.

The microphones and headsets in our helmets allowed Bruce and me to maintain a running conversation with the Lockheed and Johnson experts monitoring the test and the Marshall engineers overseeing it. Some members of the Marshall team had worked on spacewalks since the days of Skylab, the first US space station, and had helped develop the first complex orbital (as opposed to lunar) spacewalks ever done. They were full partners in the effort to develop this novel foot restraint.

Bruce and I got into the water and set about proving to any remaining doubters that the shuttle foot restraint would be useless on Hubble. Its design presumed that the axis of the socket was perpendicular to the structure you needed to work on; installing it was like sticking an electrical plug into a wall socket. We needed a work platform that would put our bodies parallel to the telescope, so that we could reach inside it for maintenance. Fisher had realized this, and so had built the mounting arm on his prototype at right angles to the hexagonal pin. The first idea that emerged during our tests was to make this bend adjustable, instead of fixed at a ninety-degree angle. This would give better access to fittings at the very top or bottom of an equipment bay and ensure that astronauts of different heights could do the work. Everyone saw the virtue of this, but nobody could imagine the mechanism that would let an astronaut make this adjustment while still in the foot restraint. We put that question aside and continued testing the platform at different worksites. Another discovery was that the six mounting positions provided by the hexagonal pin and sockets were not enough. With mounting positions sixty degrees apart, we were too often forced to choose between a position that jammed us so close to Hubble that we could barely move our arms or one that put the telescope out of reach. The obvious answer was to make the sockets twelve-sided, to allow finer-scale adjustments. The EVA engineers made a note to investigate whether this was feasible with the metals and manufacturing techniques involved.

The foot pedals Fisher had included in his design worked pretty well. It was easy to slip one boot out of the foot restraint, and the suit allowed enough lateral leg movement to get the toe of a boot out to the edge of the foot plate. The quick and easy flexibility this offered was a big step forward. On the downside, the pedals were too narrow to hit reliably; wider would be better. My recollection is that the foot plate on the prototype we tested could only rotate about one axis, either pitch or yaw, and that both foot pedals had the same function. The two pedals allowed an astronaut to use either foot, not to make adjustments in two directions. Bruce and I, thinking back to our work on the real Hubble and the realities of handling many tools during an EVA, urged the designers to put another joint inside the foot restraint housing and link each foot pedal to a different rotation.

That was a lot to digest from one day's work in the water tank. We would have a much more versatile and efficient portable foot restraint if the designers could incorporate those suggestions into a new design. Bruce was not done, however. Never one to settle for a decent design when an optimal one was within reach, he turned his attention next to the shaft between the foot platform and the hexagonal pin. That should be adjustable as well. He verbally sketched a way to build the shaft out of two segments, with one able to telescope four to six inches in length and the other, closest to the foot platform, able to rotate a full 360 degrees. I could imagine the mixture of respect for his inventive genius and weary exasperation that pervaded the test control room as the engineers wrote down yet another item they had to try to accommodate in the new design.

The EVA engineers and tool designers started working on these ideas as soon as we all returned to Houston. They would work on two parallel tracks in the months ahead. The first would be to develop a prototype with the new features as soon as possible, so it could be tested in the water tank and essential refinements identified. The second would be rigorously analytical. What metals, springs, hinges, and other parts would work on this spacebound device? How strong did each part and the entire assembly need to be to bear the loads a suited astronaut would impose on it? The new device was sure to be beefier than the one designers imagined when they installed the foot restraint sockets on the telescope. Would it overload the sockets? Did the structure behind them need to be strengthened? Was that still possible, with

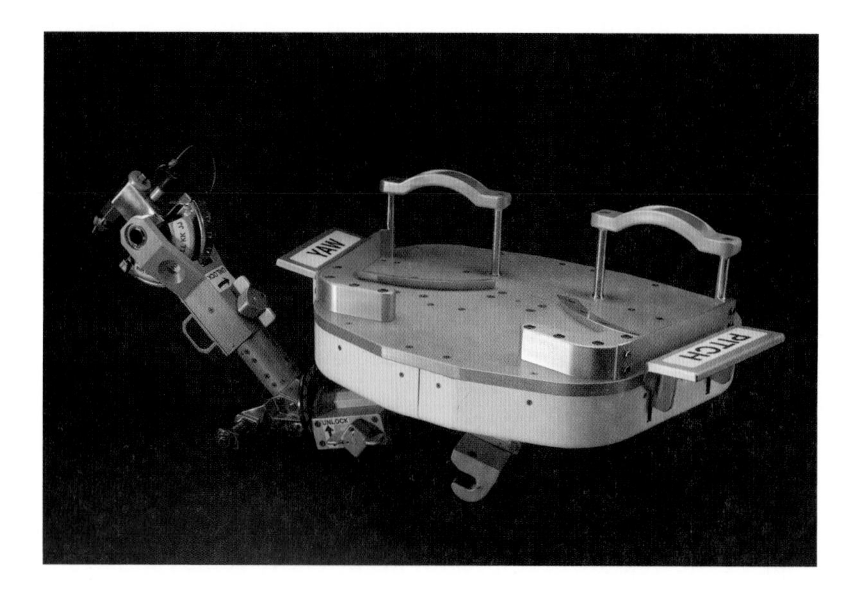

FIGURE 5.7

The Articulating Portable Foot Restraint (APFR) developed initially for Hubble Space Telescope maintenance tasks and, with minor later modifications, in use on the International Space Station. Photo by Mark Avino. Source: National Air and Space Museum.

the telescope fully assembled? These two tracks would advance simultaneously and intersect periodically, sometimes meshing smoothly and other times causing fierce debates or a design change, until a final design was produced, reviewed intensively and approved for addition to the Hubble inventory (figure 5.7).

✷

The new adjustable foot restraint was sure to be a big time-saver for future Hubble servicing crews, but the extra flexibility came with a size and weight penalty that posed a new problem for our potential deployment mission EVAs. The foot platform was now a seven-inch-thick slab that measured twenty inches by sixteen inches across. Add in the positioning arms, and the whole assembly was nearly three feet long, weighing forty-five pounds. On a maintenance mission, Hubble would be anchored to a platform in the cargo bay, and an astronaut riding the robotic arm like a cherry picker would install the foot restraint wherever needed; its bulk and heft would hardly matter. We could not use the arm this way on our flight, because it would be holding the telescope. Bruce and I would have to move hand over hand to our worksites, carrying all our tools and this bulky new device to boot. Holding it with one hand while moving along to the worksite with the other hand was both unsafe and physically impossible, since some handrails on Hubble were a full arm span apart. Attaching it to ourselves with a standard fabric tether was also out of the question. A forty-five-pound mass wafting around on the end of a tether, propelled by every movement of our bodies, was too likely to damage the telescope. We needed a way to attach the foot restraint firmly to our space suit and to put it behind us, so we would have both hands free to maneuver around on the telescope. What kind of device would be stiff enough to hold the foot restraint behind us while we moved but flexible enough to pull back into view when we were ready to install it? As usual, most of the ideas we batted around came from familiar gadgets and prior experiences. I suggested that we needed something like my flexible camera tripod, made of linked balls that could be twisted into many shapes. Robert Trevino chimed in with the news (to me, at least) that the EVA tool designers had come up with a similar concept several years back, but never developed it fully. All agreed that it might be a good starting point for the new device we needed.

The next problem was how to attach this new gadget solidly to the spacesuit. Two conical sockets near the waist of the suit seemed the obvious answer, but they existed to carry the small tool rack known as the mini-workstation. Was there a way to mount both devices to those sockets? It took Bruce scarcely five minutes to envision a way to solve this problem and produce a rough sketch of the mounting plate design he had in mind.

Armed with their insights from the water tank test, my tripod idea, and Bruce's sketch, the EVA tools crew set to work. Within a matter of weeks, Bruce and I took a prototype device into the water tank and confirmed that this "semi-rigid tether" was a very workable way to transport large objects like the foot restraint when the robotic arm was not available. Each of our spacesuits would be equipped with one for the deployment mission. The semi-rigid tether, like the new Hubble foot restraint, would prove so useful that it would become standard equipment in the shuttle and International Space Station inventories. Today's astronauts use it as a "body-restraint tether" to hold themselves firmly in position when they need both hands free to manage tools or do simple tasks on the outside of the space station.

*

The closing months of 1985 were a happy time for me. Getting back into the water for the neutral buoyancy simulations in October and November had been a real treat, one that marked a welcome shifting of gears. Not only was the pace of Hubble activities increasing, but the space policy commission that had consumed so much of my time during the prior year was winding down. It would end altogether with the release of our report in late January. The first batch of new EVA tools would be ready early in the New Year, and I would fit-check them on Hubble during the last week of January.

Or so I thought.

6

GROUNDED

I was able to catch an early morning flight out of San Jose on January 28, 1986. Getting back to Houston in the early afternoon would let me spend a few hours in my office, clearing the travel-induced backlog of mail and paperwork that no doubt awaited me. I was so exhausted from the three long nights we had worked in the VATA that I fell asleep the moment my rear end hit the seat cushion and didn't wake up until we touched down in Dallas around noon.

Walking to the gate for my connecting flight to Houston, I realized that I was still far too tired to get any useful work done that day. I stopped at a phone booth to call the office to let my secretary know I would not be coming in after all. I didn't notice anything unusual when Jessie Gilmore answered, so I said hello and proceeded to tell her that I now planned to head directly home when I landed. The odd pause that followed was my first clue that something was amiss. The quavering tone of her next words signaled that very bad news would follow. "Didn't you hear what happened?" she asked. "The shuttle blew up." I don't know if my mind froze or went blank or both, but the world seemed to stop completely. I could not make any sense of her words. "The shuttle blew up" was too abrupt to fit any disaster I could imagine. A sentence like "We aborted because of an engine failure and the shuttle crashed short of the runway" would have made sense, despite the equally tragic result. "The shuttle blew up" just did not, could not compute.

I boarded my connecting flight in a numb fog. A space shuttle had blown to bits, taking with it seven people, including four of my TFNG classmates. Reporters occupied every seat surrounding me, eight of the many who were racing to Houston to grab a piece of this career-making story. Luckily, I was not wearing any NASA logo gear and could remain anonymous. One guy was replaying the explosion over and over again, presumably on his camera's video monitor, and gushing about the scene as if he were critiquing the visual effects in an action movie. They were clearly thrilled about being in on such a huge story, so thrilled that they barely managed to remind themselves once or twice that seven people were dead. I wanted to scream a thousand things at them and make them shut up, but knew an angry outburst would only make things worse for both me and the agency.

I headed directly to the astronaut office after landing, hoping to find some reliable information about what had happened and knowing that the only people who understood what I was feeling would be there. As if to confirm that the world was now utterly, irrevocably altered, the space center itself seemed to look different as I drove through the gate. I felt as though someone had paused the movie I was in, but I was somehow still moving through the frozen, silent scenery. A leaden silence permeated the empty hallways of the astronaut office spaces on the third floor of Building 4. Much of the astronaut corps had been out of the office that day, either supporting the launch or on technical travel, as I had been. Those who were in Houston had rushed home as soon as possible after the accident to be with their families. Jessie and the rest of our secretarial cadre had remained on duty. Red-eyed and teary, they were gamely fielding the flood of calls that was pouring into the office from astronauts and spouses and all the reporters hoping to get a dramatic sound bite or bit of inside information. I imagine I hung around the office for several hours before heading home, but all I really remember of that afternoon are the vacant stares and tear-stained eyes that surrounded me, no doubt mirror images of my own face, and the stark sense that my entire world was in free fall.

The next several weeks would be a surreal and trying mix of grieving and caring for the families of the lost crew—delivering meals, helping out with the kids, wiping away tears, and fielding anguished questions for which there weren't yet any answers. Many evenings found me talking late into the night with June Scobee, the newly widowed wife of *Challenger* commander Dick Scobee, at the butcher-block table in her kitchen. Brainstorming with her about what kind of educational initiative could be a living legacy for the crew was the only therapy I could offer.[1]

*

Nothing got easier about 1986 as the months went by. The presidential accident investigation seemed to slog along, with each week bringing more shocking revelations about lapses in the technical competency and integrity of the NASA team I had trusted with my life.[2] No good answers about the way forward accompanied these revelations. Was the accident caused by a fixable problem or a fatal design flaw? Would we ever fly again? It seemed nobody knew, nor had any idea when we might know the answers to these and countless other questions that would dictate our future. We toiled on despite the uncertainty, trying to maintain some sense of direction, but the atmosphere in the astronaut office and across the entire Johnson Space Center campus remained distinctly glum.

The human losses kept mounting as well. A favorite cousin died in April, and a light plane crash in May took the lives of AS-CAN Steve Thorne and one of my favorite mission control engineers. That made ten deaths in under five months for me. I was filled with sadness and anger, and frustrated that there was nothing I could do to help us get back into space.

One of the people I commiserated with frequently was shuttle instructor pilot Charlie Justiz, a sympathetic soul with a zany sense of humor who is the friend and confidant of many astronauts. The

FIGURE 6.1

The crew of *Challenger* mission 51L in the launch pad white room after completing their countdown dress rehearsal. Left to right: Christa McAuliffe, Greg Jarvis, Judy Resnik, Dick Scobee, Ron McNair, Mike Smith, Ellison Onizuka. Source: NASA.

accident had rocked his world as much as it had mine. We joked that 1986 had become such a horrible year that we should just declare it to be done—throw it out like you would a bottle of bad wine—and start a new one. That gave me a good laugh, and Charlie an idea about how he could help raise people's spirits: throw a New Year's party, complete with a ceremonial moment to mark the end of the old year and start of the new. The party at his lakeside home in July was the first time that year that we had gathered to laugh and enjoy the company of friends rather than mourn our losses. At the mock New Year's Eve moment, we raised a toast to our lost crewmates (figure 6.1), cursed 1986, and wished each other a happier New Year.

*

As it became clear that the fleet would be grounded indefinitely, many shuttle payload customers began to look for other ways to get into orbit. The commercial customers NASA had courted so ambitiously were the first to defect, shifting their communications satellites to conventional expendable (single-use) rockets to minimize business losses. The Department of Defense, eager to follow suit, lobbied furiously to reverse the national policy that had forced them to use the shuttle as the primary military launch vehicle. Customers with payloads that were designed specifically for launch on the shuttle were stuck with hoping the hiatus would not be too long and waiting to see where they would end up in the cargo manifest when flights resumed. The missions locked into waiting for the shuttle included a few classified military payloads and commercial satellites, two NASA tracking and data-relay satellites, two NASA planetary probes, and Hubble. NASA would clearly give top priority to national security and paying customers when the time came to develop a new flight schedule. The planetary probes would be next in priority, because missing their narrow launch windows would trigger delays of many years. Sadly for Team Hubble, no such external factors existed to force NASA's hand in rescheduling us. The telescope was bound to slip far down in the new launch sequence. Our wait was sure to be measured in years, not months.

Hubble activities had continued as originally planned through February and March, as everyone absorbed the scale of the technical and programmatic challenges that would have to be overcome before the shuttle could return to flight. The new reality—a long delay of unknown duration—quickly catapulted the cost of maintaining the telescope's technical workforce to the top of the list of problems facing the program managers. Hubble had been just nine months from launch before the accident. NASA and all the contractor companies had staked their budget plans and profit outlooks on the reduced staffing levels and running costs that putting the telescope into orbit would bring. Now they faced yet another schedule slip and more unplanned costs. It hardly mattered that this slip and budget overrun were not of their doing. Containing costs during the shuttle hiatus, however long that would turn out to be, was understandably the top concern for senior Hubble managers in Washington, Huntsville, and Sunnyvale.

The Hubble M&R team saw the hiatus as the opportunity to complete the maintenance upgrade work we had begun back in mid-1985. At that time, we feared that the few months remaining before Hubble would be shipped to Kennedy Space Center for our October 1986 launch would prevent us from getting much done. The small bit of silver lining we saw in the *Challenger* accident was that the fallout from it would allow us to make the telescope fully maintainable and guarantee that the tools future astronauts took to orbit would work properly. We were bound and determined that the words "Hey guys, it doesn't fit" would never be uttered on a Hubble servicing mission.

Bruce McCandless was the ringleader of our campaign to use the hiatus in this fashion. He was aided and abetted very ably by Ron Sheffield, who now led Lockheed's Hubble maintenance and repair team. Bruce seemed to know everybody who was involved with Hubble in every organization, from the senior officials at NASA headquarters to the technical teams at Marshall, Johnson, and Goddard and the engineers working for Lockheed and the subcontractor companies. More importantly, he had an astute grasp of the organizational dynamics and power relationships involved. These would prove indispensable to our campaign in the years ahead, as we collaborated and conspired with Sheffield to win support for the changes needed to expand Hubble's maintainability and lay the groundwork for future servicing missions.

*

At the time of the *Challenger* accident, just fourteen systems were designated as "on-orbit maintainable" under Lockheed's contract with NASA. The number of individual modules in each system varied from one to twelve, bringing the total count of individual units designed for EVA maintenance up to forty-three. These were now designated as "Block I On-Orbit Replaceable Units" or ORUs. Lockheed's 1985 reassessment of the criticality and reliability of the telescope's systems had produced a prioritized list of additional units to modify for on-orbit replacement. These fell into two batches, with twenty-six systems and fifty-one units in "Block II" and a further eight systems and sixteen units in "Block III." If we could push through both the Block II and III lists, Hubble would be roughly twice as maintainable as before.

Sheffield scheduled a day in September for Bruce and me to come out to Sunnyvale to do a top-to-bottom inspection of the Block II and Block III units on the telescope and review his team's initial ideas about how to make them maintainable on-orbit. The idea of making any modifications had met with strong resistance in some quarters. The telescope was fully assembled and well along in its battery of preshipment tests at this point. Access to the VATA and the work platforms surrounding the telescope was limited strictly to essential engineering and test personnel to reduce the risk of inadvertent damage. Allowing more people near the telescope just increased the hazard level. Our first battle was to convince the powers that be that we belonged in the essential personnel category rather than the greater hazard category. The second point of resistance was hypothetical: the push to modify components for EVA compatibility might open the door to disconnecting and removing hardware that had already been installed and tested successfully. Some people, notably Frank Costa, the czar of all things electrical, understandably preferred to keep this door nailed firmly shut. It took a pincer attack by Ron and Bruce, arguing up both the Lockheed and NASA chains of command, to overcome these obstacles and get us cleared for work on the telescope.

In addition to Ron, Bruce, and me, the team suiting up to evaluate the Block I and Block II modules included Peter Leung and Brian Woodworth from the Lockheed M&R team and one of the EVA experts from the mission control team in Houston.[3] As we made our way down the telescope, we would be met at each stop by an engineer from the team responsible for the gear inside that equipment bay. Frank would be with us every step of the way, of course, both sharing his encyclopedic knowledge and guarding his telescope like the fiercest of mother hens.

It took only a glance to spot some of the differences between components that had been designed for EVA replacement and those that had not. The arrangement and design of the electrical connectors jumped right out at us, for example. Connectors on the boxes designated as maintainable were spaced far enough apart to allow an astronaut's gloved hand to access and operate them. They also had a backshell, a sheath of metal extending several inches back from the connector, that prevented damage to the individual wires during maintenance operations. The rotating collars or rings that locked and unlocked connectors on the EVA-compatible boxes had wing tabs to make it easier for spacewalking astronauts (as well as ground technicians) to connect and disconnect them (figure 6.2). The Block II and III units lacked all of these features.

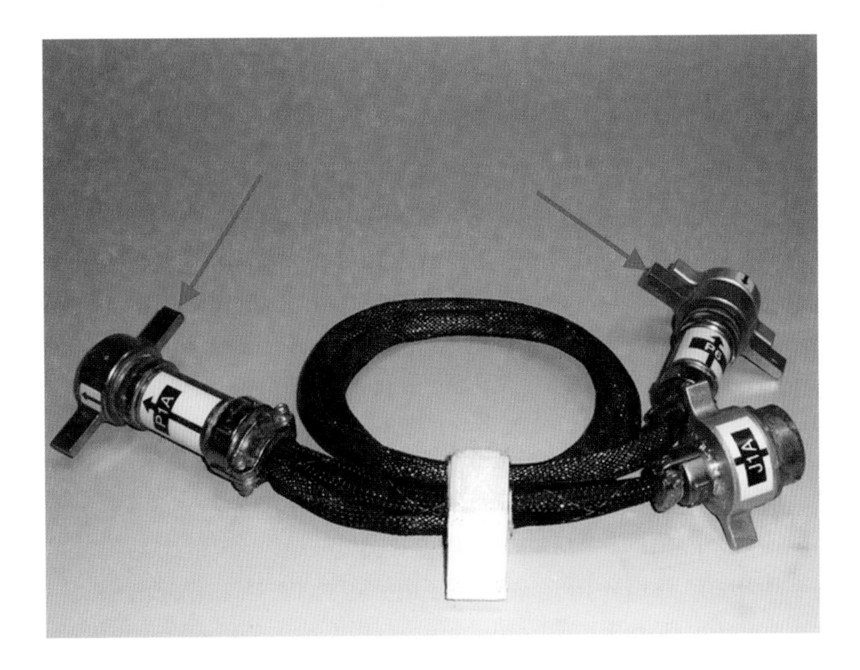

FIGURE 6.2

Electrical connectors with wing tabs (red arrows). Source: NASA.

The next obvious and important thing to check was how the candidate boxes were mounted to the telescope. The standard EVA fastener on Hubble was the same one used on the space shuttle: a 7/16-inch hexagonal-headed bolt with a double-height head. Use of this fastener was not the only feature of importance for EVA maintenance, however. Every skilled maintainer knows how critical it is to keep track of all the small components she or he has to handle in the course of a task, like nuts, bolts, and washers. Sorting them into jars and bins or arranging them on strips of two-sided tape are common ways to avoid losing any of these on Earth. None of those tricks works in a zero-gravity environment. As a result, all the fasteners on Hubble's EVA-compatible boxes were captive, meaning an astronaut could loosen the bolts to remove a module without worrying about any of them floating away. In some cases, slots shaped like keyholes were machined into the mounting flanges. The wider end of the slot made it easy to align the module on all the bolts simultaneously during reinstallation. Many of the modules we now hoped to make replaceable on orbit lacked these features as well.

The final point we evaluated for each candidate module during our inspection was worksite accessibility. What EVA handrail route led to the equipment bay in question? Where would be the best place to mount a portable foot restraint for this task? Was the prototype Hubble foot restraint then in the inventory adjustable enough to meet our needs? For each potential new task, we asked ourselves whether anything appeared so tricky that it should be evaluated more carefully by a test on the Hubble mockup in the neutral buoyancy training facility.

It took several long sessions in the VATA to inspect Hubble thoroughly from head-to-toe, but we finished with a comprehensive inventory of the modifications needed to make the Block II and Block III units either fully EVA compatible or, failing that, at least more EVA friendly.

As we had expected, some of the needed changes would be quite simple to implement and others much more complex. The simplest item on our list was the addition of labels to each connector on every module. Bruce and I knew from our previous EVA experience how vital these would be to servicing mission astronauts. Plugging the wrong cable into a socket or cross-threading a connector because you couldn't see or feel the proper alignment could spell disaster for the telescope, and both could happen all too easily, given the limited visibility and dexterity a spacesuit provides. Even this simple idea encountered resistance, though. It would cost money and take time. Worse, it was low-value activity that needlessly exposed the telescope to extra risk of damage due to human error; the fewer people who touched Hubble, the better. Some of the telescope engineers went so far as to ridicule the idea. They hadn't needed any such aids to assemble Hubble, they argued, and besides, the labels would probably peel off before any astronaut arrived to repair the telescope. Only some of them were teasing; others seemed intent on avoiding what they saw as a risky nuisance.

The electrical switching station known as the Power Control Unit or PCU—the very heart of the Hubble Space Telescope—posed the greatest challenge. All the electricity produced by the solar arrays feeds into this unit, which then distributes it to the telescope's basic operating systems and scientific instruments. Frank Costa tried to talk us out of even inspecting it. He said it was impossible to replace the PCU in flight because it could never be powered down. Dozens of reasons why replacing it might be impossible jumped out at me as I looked into the equipment bay, but none had anything to do with his objection. With deliberate theatricality, I shot back that if a failure in the PCU disabled an otherwise healthy Hubble Space Telescope, he could bet his sweet life that NASA would mount a mission to replace it. Our job was to identify modifications that would make this more feasible than it was at present. Making the PCU fully compatible with EVA servicing was clearly out of the question, but we had to find a way to move it from what I called the "Utterly Impossible" list to the "Horrendously Difficult" list.

Honestly, though, my first thought when I looked at the PCU inside equipment bay 4 was "no way" (figure 6.3). For starters, the door to this equipment bay could not open 180 degrees like all the others. The thick cable bundles that ran along the hinge line limited it to just half that. Compounding this, four electronics boxes mounted on the door's inner surface effectively made the door eight inches thick. Even with the freedom of motion provided by our bunny suits and the VATA's large work platforms, it was tough to reach inside that bay, and impossible to get a clear view of the left-hand side of the PCU. These reach and access problems alone seemed a show-stopper at first.

Then there was the PCU module itself. This huge box almost filled the bay entirely, and thick bundles of copper-colored cables packed what little bits of space it did not take up. A dense forest of electrical connectors, thirty-four in all, jutted out from one side. Of course, this was the left side, the one closest to the door hinge, with the worst reach and visibility problems. The connectors were spaced so closely together that I could barely get my fingers around the closest ones, and the collars that locked them onto the box were too tight to release with just fingertip force. I dared not try a full-handed grip, though, because of the dozens of fine wires that ran into the back of each one. Like the arteries leading in and out of your heart, every one of these wires carried a vital power or feedback circuit. No backshell protected them from the strain I would surely impose with a full-handed grip, and I knew it wouldn't take much strain to dislodge the wire and break a vital circuit.

FIGURE 6.3

Equipment bay 4, showing the original installation of Hubble's Power Control Unit (large black box inside the bay) and the tight work area caused by the Power Distribution Units mounted on the door (left) and cable bundles that limited the door's opening range. Source: NASA.

Next came the mounting bolts. The five bolts I saw on each side of the PCU were not EVA standard, but at first glance I thought they might be manageable. That wishful impression vanished when Sheffield gave us three additional facts. First, there were actually ten bolts to a side. The other five were inserted from the back side of the PCU's flanges. On top of being invisible from the front of the module, these bolts were covered by thermal insulation blankets. In the VATA, it took multiple sets of hands to push the thermal blankets out of the way, hold the mirror that was needed to see the bolts, and turn the wrench. Second, none of the bolts was captive and, third, each had a noncaptive washer on it. In the VATA, these small items would drop into your hand or fall to the bottom of the equipment bay when they came loose. In the microgravity environment of the shuttle they would float around, easily propelled out of reach by the inadvertent brush of a tool or a hand and horribly hard to capture wearing a bulky spacesuit glove. More bad news followed. Three electrical grounding wires were secured to a single bolt on the bottom of the module with a noncaptive nut. We needed an inspection mirror and very nimble fingers to get the lugs on the end of each wire onto the bolt and hold the trio in place while engaging the nut. All these factors made installing or removing the PCU extremely maintenance-unfriendly. Even for the bunny-suited technicians in the VATA, it was at least a three-handed proposition. The reach and visibility limitations imposed by a spacesuit would clearly make it impossible.

A lot of "what if" and "maybe we could" notions were floated as Bruce and I took turns studying the PCU. With our inspection done, we gathered on the work platform to discuss what, if any, solid options we could come up with. The mounting bolts were clearly the go–no go factor: unless we could find a way to make those EVA accessible, it would be impossible to replace the PCU in orbit. The trick was how to do this on an already-built telescope, and with the lowest possible added cost or risk of collateral damage. The breakthrough for me came when I probed again how the PCU was installed. Was it mounted directly to the inner surface of the equipment bay, what we called the tunnel wall? It was not. Instead, it was mounted onto a bracket, which was in turn attached to the tunnel wall. Suddenly the problem struck me as being equivalent to traveling internationally with electrical devices. All it took for the American plug on my gadget to connect with the sockets overseas was a simple adapter with a US socket on one side and the foreign country's plug on the other. We just needed to insert a plate between the bracket and the PCU that would match with the bracket's non-EVA fittings on the back and provide standard, captive EVA bolts on the front.

The five of us—Bruce, Ron, Peter, Brian, and me—kicked this idea around for perhaps thirty minutes as we stood by the PCU. The adapter plate would have to be strong enough to handle the harsh vibrations and accelerations of launch. The plate would actually worsen these by moving the mass of the unit a bit further from the telescope's central axis. Conversely, it would have to be so thin that the equipment bay door could still close. Was there a design that could meet these conflicting requirements? What about thermal issues? Would the adapter put the PCU out of its specified operating temperature range? If the idea proved feasible, were there other features we could add to make the unit more EVA friendly? Shifting the PCU a bit to the right to make it a little easier to work on the electrical connectors seemed like a good idea, as did finding an alternative design for the grounding wires.

In this case, Bruce's and my creative involvement effectively ended with the conclusion of this brainstorming session. Our time had to be spread across too many issues and activities to allow us to plunge into the iterative engineering design process that would ensue. From this point on, we would periodically review the modifications that Sheffield and his team came up with and throw our weight behind them as they went forward to Lockheed and Marshall management for approval. Building on the ideas we batted around in the VATA, Ron's crew developed a preliminary design for a two-piece adapter. One piece would attach to the existing mounting bracket and the second to the back of the PCU itself. Both would be several inches wider than the existing flanges on the PCU, creating space for ten J-hooks like the ones used to latch the equipment bays closed. Adopting this design would solve the invisible bolt problem and, helpfully, shift the PCU module an inch or so to the right. The challenge of handling three grounding wires on the bottom of the module would be solved by attaching all of them to a single metal plate and moving the mounting bolt to the front.

Sheffield's first attempt to get the NASA's approval to invest time and money in the engineering analysis needed to start turning this concept into real hardware was rejected. Undeterred, he directed Brian Woodworth to write a detailed memo to Tom Dougherty, rebutting the review panel's decision and including pictures of the key problem areas and the measurements they had made on Hubble to prove the adapter would fit. Then he called us for help. Bruce was not one to waste time debating an important issue with someone he believed had made a bad decision when he could appeal directly to a higher authority instead. He called the senior Hubble program manager at NASA headquarters and demanded that he direct Marshall to approve the PCU modifications. This led to a cascade of phone calls from headquarters to Marshall to Lockheed that ended with Bert Bulkin bellowing, "I'm going to fire you!" as he strode down the hallway to berate Sheffield for going over his head. Ron survived, as he would on many other occasions in the years before launch, and work began to design and produce the PCU adapter plates that would save Hubble's life in 2002.

✷

The dense forest of large electrical connectors on the left side of the PCU forced another innovation. No tool in either the Hubble or shuttle inventories at the time could grab such tightly spaced connectors. Worksite geometry posed a second problem the tool would have to deal with. Many connectors were hard to see and awkward to reach because of their placement in the equipment bay. These reach and visibility factors, plus very stiff locking rings, made them difficult to connect or disconnect, even in the VATA. For a spacesuited astronaut anchored to one of the foot-restraint locations on the telescope, operating these connectors was sure to be a semiblind, one-handed operation. So, we needed a single-handed tool with skinny fingers that could get a very firm grip on the connectors without putting strain on the cables. A tool that gripped the connector rings from the side, like pliers gripping a nut, would be safest for the wire bundles, but the cramped spacing would allow it to turn only a few degrees. The new tool we needed would have to grab the connectors from the top instead.

This challenge was right up Michael Withey's alley. Withey was a member of the ILC Space Systems team that produced and managed all the tools and equipment that shuttle astronauts needed to have in flight. Their work ranged from initial design through fabrication to logistics and onboard stowage, and covered items as varied as EVA tools and spacesuits, meal trays, and laundry bags. The son and grandson of diplomats, Michael had lived in Chile, Turkey, and Ethiopia before his family settled in Brownsville, Texas, for his high school years. With a passion for motorcycles and an aptitude for all things mechanical, he went to South

Texas Junior College, hoping to become an engineer. In 1966, one of his professors told Michael he should go to Houston and get a job as a draftsman, if he was serious about that goal. He landed a job with Lockheed Electronics at the Johnson Space Center, but had to put his career plans on hold when he was drafted in 1968 for a two-year hitch in the Army. He went back to Lockheed when he returned to Houston, then moved to ILC in 1979, quickly becoming the anchor of the company's EVA equipment team.

As usual, Withey scarcely looked at Bruce or me as we talked through the Hubble connector challenge and our initial thoughts about the kind of tool we needed. He doodled on a sheet of paper while we talked, as if we were boring him. We knew better than to think that, though. Michael was transforming our snippets of thought into a detailed initial design drawing as we spoke. He drew a pliers-like device with thin metal bars mounted at right angles to the shaft where the jaws would usually be. These were the slender fingers needed to fit into the tight spaces between the connectors. He had converted our description of the high torque required to turn the locking rings into first estimates of the best metal to use, the dimensions and fasteners needed to attach the jaw extenders onto the shaft, and the torque level the tool could provide. The three of us talked through these points and other features it needed to have. Bruce and I worried about using a tool with metal jaws on the Hubble connectors. If the tool slipped while an astronaut was gripping it tightly, it might scrape small flecks of metal off the connector rings. Creating metal debris that could float among the electronics modules in an equipment bay was clearly a bad idea. The jaws would have to be lined with some nonmetallic material. This needed to be soft enough to grip the connector ring but not so soft that it would be destroyed by the high forces involved. What kind of plastic or rubber would fit the bill? The next problem was how to add this material to the jaws. Could it be applied as a coating? Glued on? Could there be a small cavity on the inside of each jaw to hold a chunk of the material?

Michael set off to build a prototype of this new tool, based on his initial design and the points we had debated together. When it was finished some weeks later, the three of us took it out to Sunnyvale for testing.[4] The only good news was that the jaw extenders were indeed thin enough to get in between the cables and onto the connectors; little else worked as needed. When we squeezed the handles to twist a locking ring, the top of the tool closed tightly, but the extenders splayed out like the legs of a wobbly newborn horse. The force we applied to the handle was not being transmitted down to the locking ring. There were two problems here. The extenders needed to be attached more strongly and the simple pliers-like design did not keep them parallel as force was applied. There were problems with the jaw liners, too. The material was too hard to grip the connectors well and, even with the weakened grip caused by the splaying jaws, small pieces of it were dislodged. Again, we brainstormed possible solutions to these problems while still alongside the telescope, so we could recheck details as our evolving ideas triggered yet other questions. Sheffield's team always took part in these tests, adding their creative ideas and knowledge of the telescope to the mix.

Unsurprisingly, Withey could see the new hinge mechanism and other fixes we needed in his mind's eye before we left the VATA. The revised version we tested a few months later solved all these problems and raised no new ones. With a blessing from Bruce and me, Withey and Sheffield began the engineering and administrative processes needed to turn the new design into a flight-certified tool and officially add it to the Hubble maintenance toolkit (figure 6.4).

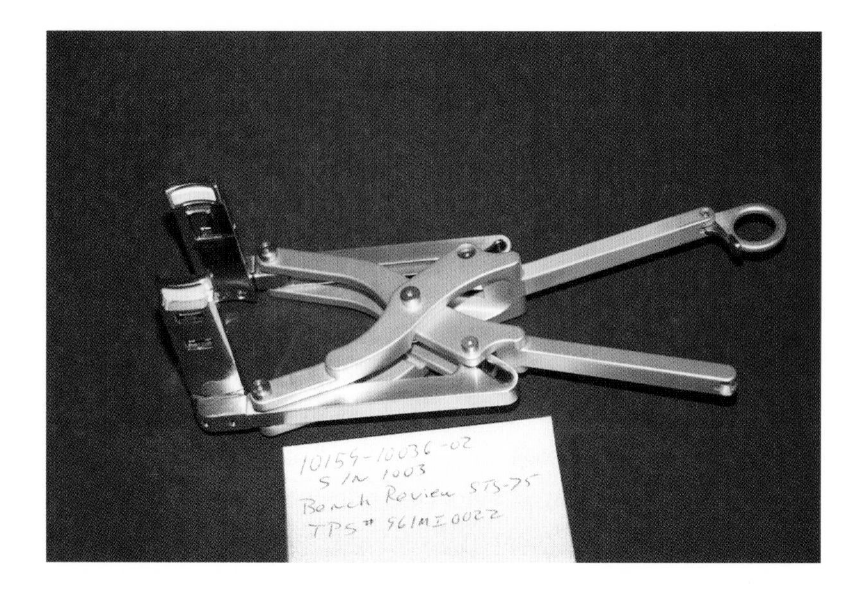

FIGURE 6.4

Final flight version of the 90-degree connector tool. Source: NASA.

The question of how to manage the "downtime" or "stand-down" was a major topic at the quarterly Hubble program meeting in August 1986, even though we still had no idea how long this hiatus would last. The plan was to keep up the pre-accident pace until all flight elements were installed, all testing was complete, and Hubble was ready to ship to the launch site. Everyone would then cut back to a maintenance level of effort. Once a new Hubble launch date seemed reasonably firm, we would review every team's restart plan and ensure that all the work streams synched up correctly to meet that date.

Our bosses in Houston had also been wrestling with this question. Costs worried them too, of course, but an equal or greater concern was how to prevent the flight operations team—astronauts and mission control engineers alike—from going completely stale during an indefinite hiatus. One key to this was to keep people working toward a target date, even if it was only a wild guesstimate or an altogether arbitrary target. Another was to keep folks busy. With Hubble work sure to slow down to maintenance mode, they felt it no longer made sense to have two astronauts assigned to it full time. Having already endured long periods of thumb twiddling, Bruce and I could not have agreed more. The powers that be decided that Bruce would remain the day-to-day Hubble guy and my primary job would shift to serving as technical advisor for half a dozen payloads that were in the future cargo pipeline (only two of which would eventually fly). I would keep my hand in on Hubble issues and take part in major tests to the extent that time outside of these duties allowed.

This new assignment would fill my normal working hours pretty well, but without launch preparations or real-time flight operations going on in the background, life still felt much slower than usual. I managed to attend most of the quarterly Hubble status meetings and many meetings of the newly established Maintenance and Repair panel. In my "abundant spare time" (a common bit of astronaut sarcasm), I continued helping June Scobee with the launch of her educational initiative, now called the Challenger Center for Space Science Education, began pursuing my multiengine pilot rating, and parlayed some old oceanographic connections into an opportunity to join a research cruise. Aboard ship, I met a fascinating former Navy oceanographer by the name of Andreas Rechnitzer. Andy had many great tales to tell about the exciting projects he had worked on during his Navy career, most notably the famous 1960 dive of the bathyscaphe *Trieste* to the deepest point in the ocean. This and other stories he shared during our late-night conversations started me thinking again about whether joining the Navy Reserve might be a good way to keep my hand in oceanography. Each time I had explored this notion in years past, I had been warned to steer clear, on the grounds that the Navy would only make me somebody's secretary. Oceanography had become a distinct professional path within the Navy Reserve since my last inquiry, however, and I wondered if that would change the answer. I decided to talk to Bruce about it when I got back to the office.

If Bruce seemed to know everybody of importance in the Hubble universe, this was triply true in the Navy, owing to his long service and family heritage as the son and grandson of two admirals, each of whom had been awarded the Medal of Honor. One October day, a few weeks after I returned from the research expedition, I remembered to ask his advice on how to explore the Navy Reserve prospect. Peering at me over the reading glasses perched on the end of his nose, he said nonchalantly, "Well, I'm having lunch tomorrow with the Secretary of the Navy. Would you like me to ask him?" "Sure," I said, with as much nonchalance as I could muster, "by all means, please do." Thanks to Bruce's lunchtime intercession the following day, I would become a Lieutenant Commander and begin my own Navy adventure before another space shuttle left planet Earth.

*

Another job was added to my plate two months later: Capcom (short for Capsule Communicator) for the first three missions when we eventually returned to flight. From Project Mercury through the space shuttle years, the astronaut serving as Capcom was the sole vocal link between the crew in orbit and the mission control team on the ground. It's a coveted assignment—the second-best job an astronaut can have. I was thrilled to have a part in the return-to-flight effort, no matter that nobody knew when this would begin in earnest.

7

RETURNING TO FLIGHT

In January 1987, NASA announced the five veteran astronauts who would fly STS-26, the first post-*Challenger* shuttle flight. My TFNG classmate Rick Hauck would command the mission, with fellow TFNG Dick Covey as his pilot. The three mission specialists for the flight would be George "Pinky" Nelson, another TFNG, plus Dave Hilmers and Mike Lounge from the Class of 1980. As expected, the primary payload for this first mission would be another tracking and data-relay satellite, to replace the one lost with *Challenger*. Two military missions came after this flight in the new manifest, followed by the Magellan probe to Venus, a NASA Spacelab mission, and then the Hubble Space Telescope.

The crew announcement struck me as a "stake in the sand" move on NASA's part, an effort to inject some focus and one element of certainty into what was still a very fluid situation. The return-to-flight effort was geared toward an early February 1988 launch, despite ample evidence that the amount of technical work yet to be done would make this date impossible. Each of the nine recommendations made by the accident inquiry board, known as the Rogers Commission, had cascaded into dozens of technical and managerial actions that had to be completed before the shuttle could fly again.[1] One of the very long poles in the tent was the development of a system that would allow a crew to escape from a damaged vehicle. Two options were being evaluated—tractor rockets to pull the crew away from danger (à la Apollo) or bailout during gliding flight. Testing and analyses were still underway to determine which of these options was safer and more technically feasible. There was no way either system could be selected, built, and installed by the supposed launch date. Another pacing item was the technical reassessment of every critical equipment component built into a space shuttle. This included detailed analysis of the ways in which each one could fail and what remedial measures would reduce the likelihood or consequence of a failure. Chief Astronaut John Young announced at one of our weekly All Astronauts meetings that he counted 201 specific technical actions that remained open, on top of all the new actions that were sure to come out of the critical items review. We still had a long way to go.

Despite the uncertainty in the launch date, I felt like the ground beneath my feet was firming up at last. It felt good to once again be working on the technical building blocks that would eventually result in a spaceflight, no matter how far off that might be. I gave my Capcom duties top priority, religiously attending all the technical reviews, planning meetings, and training sessions needed to get myself back up to flight-ready proficiency on shuttle systems and ensure that I understood the payload operations every bit as well as the flight crew. Hubble developments wove through this backdrop of activities like distinctive threads in a tapestry. Occasionally the opportunity would arise to hop into the water tank to evaluate maintenance tasks and support equipment, or to take some of our prototype tools into the VATA to confirm that they fit and functioned as planned. I jumped in eagerly whenever I could squeeze these in amid my return-to-flight duties. When I could not break free, we gave my slot to another astronaut to continue familiarizing more potential future spacewalkers with the telescope.

*

The most memorable event of 1987 for me was an extravagant party the Soviet government threw in October. The occasion was the thirtieth anniversary of the event that marks the dawn of the space age, the launch of Sputnik on October 4, 1957. Lest anyone forget which country's technical prowess had produced that feat, the Soviets had invited the Who's Who of global aerospace to Moscow for a three-day celebration, culminating in a huge party on the anniversary date. To showcase their largesse, they offered to cover all expenses for all American attendees, including round-trip travel via Aeroflot from New York to Moscow.

Every NASA astronaut, current and retired, was invited. Retired astronauts were free to accept or decline the invitation as they wished. Those of us still on the active astronaut roster could attend only if we were tapped to be part of the official United States delegation—assuming there would be a delegation. Soviet–American relations had thawed somewhat since the US boycott of the summer Olympics in 1984, but Cold War relations were always chilly. The April 1987 discovery that the newly built US embassy in Moscow was heavily bugged had strained the relationship anew, as had the later news that the Soviets had obtained highly sensitive technical information about US submarine designs. The United States could hardly skip the event, but it did not want to contribute overly much to what was obviously both a self-congratulatory party and a propaganda ploy. Whom would we send?

The powers that be eventually decided that the United States would send a very small official delegation, headed by a midlevel NASA official and including just one current astronaut—me. As usual, I could only speculate as to why I was selected for this assignment; no reason was given.

It was quickly apparent that the diplomatic equation surrounding this trip was much more complex than anything I had encountered on prior international trips. NASA plays a significant role in the United States' scientific diplomacy. Since astronauts are the agency's most visible and celebrated ambassadors, the decision to send astronauts on an overseas trip always involved careful planning with the State Department and often required approval above the NASA headquarters level. Planning for my 1985 trips to the Paris Airshow and the International Astronautical Foundation conference had given me a glimpse of the intricate diplomatic calculus involved in this, especially when American astronauts and Soviet cosmonauts were to attend the same event. But both of those events had been held on neutral ground and hosted by an apolitical organization with a broad agenda to advance global aerospace. The Sputnik anniversary event would be hosted by an avowed adversary on its own turf, and was obviously designed to advance a political propaganda agenda. The level of scrutiny and calculation involved in preparation for this trip was vastly more intense. My excitement at getting an inside glimpse of the Soviet Union battled with an unsettling sense that I was walking into a spy novel. The trepidation side of the equation went up a notch when I learned we would stop in Paris on the way over for a counterintelligence briefing from the State Department's diplomatic security experts. It went up yet another notch as I listened to their briefing.

The pilots of the Air France jet that would take our delegation into Moscow were thrilled to have an astronaut aboard, especially an American who spoke French fluently, and invited me to join them in the cockpit for the approach into Sheremetyevo International Airport. As we crossed into Soviet airspace, they handed me a map and talked me through the mandatory arrival route and procedures for flying into Moscow. One dog-leg in the route was particularly important: if we failed to turn at the right place, we would quickly be inside a Soviet air defense zone and blinking on somebody's radar screen as a potentially hostile target. Underneath the typical pilots' banter, I could feel the tension rising in the cockpit as we neared that point. At the last checkpoint before the critical turn, the copilot radioed Moscow approach control to report our position. We all expected the reply would include the instruction to turn, but it did not. On we flew for several tense and abnormally long minutes, with the pilot debating whether it was better to avoid the prohibited airspace by turning without permission or pray he could talk his way out of an unauthorized incursion into it. When Moscow at last cleared us to change course, he rolled swiftly into a left turn that I am quite sure violated Air France flying standards.

The whirlwind of events I was swept into from dawn until the wee hours of the morning over the next three days ranged from the most banal conference-style scenes—overly long speeches, facility tours, staged photos, and collegial banter over lunch or cocktails—to moments that seemed straight out of a John LeCarre novel. The penetrating stares of the babushka posted as watch keeper in the hotel hallway and the things that moved around in my room while I was out were unnerving, to say the least. I noticed my interpreter was always nervous when we talked in the presence of our driver, a man who supposedly spoke no English but always seemed extremely alert to my words. I couldn't shake the uneasy feeling that I was a target of interest in a very high-stakes game, in which the rules were secret and all the players carefully disguised.

Notwithstanding the spy novel aura of it all, there was good fun to be had. One day, cosmonaut Aleksandr Aleksandrovich "Sasha" Serebrov and I were paired up as space ambassadors to visit a long list of schools and Young Communist groups. Sasha and I had met briefly at the Paris Airshow in 1985 and established an easy rapport based on our shared interests in space science and planetary exploration. In Moscow, our round of visits began at midday and ran well into the night. As we were driving from one event to another, he asked how much of the city I had been able to see. He was aghast when I told him there was no time at all for sightseeing in my schedule and declared that he would show me the key sights on the way back to my hotel. After our last event, we sent the official driver away, hopped into Sasha's car, and set off into the night. Policemen stopped us several times as we sped around town but waved us on quickly and apologetically as soon as he flashed his cosmonaut identity card. We raced around Moscow for hours, strolling down the Arbat, stopping at notable statues and ducking into Metro stations to admire the architecture, talking all the while about our spaceflight experiences. I was exhilarated and utterly exhausted when he dropped me back at my hotel around four in the morning. It was a gloriously madcap midnight adventure.

The most cloak-and-dagger moment occurred during the next day's cocktail reception. I was chatting with a group of people when a man bumped into me. Muttering something in Russian, he shoved a cardboard tube under my right arm and vanished into the crowd. I had no idea who he was or what he had said, but his furtive manner spooked me. Had he just tucked an innocent gift under my arm, or a piece of forbidden literature he hoped I would spirit out of the country? I quickly tracked down NASA's senior international affairs official to tell him about the incident and get his advice on what to do. The obvious first step was to see what was inside the tube. This turned out to be an etching unlike any other piece of space art I had ever seen (figure 7.1). Instead of the bright, aspirational spirit common in Western space art, this piece seemed dark and mysterious. Five spacesuited figures cluster near a sphere that is partially surrounded by a fat greenish tube. The figure inside the sphere appears to be operating a control panel, while the other four are clambering around on panels and girders suggestive of a space station. Stars and galaxies pepper the surrounding blackness. An inscription penciled onto the lower border dedicated the work to me and expressed great respect for my accomplishments in space. Was my bump-and-run stranger the artist himself, Anatoly Ivanovich Veselov? What did he mean to

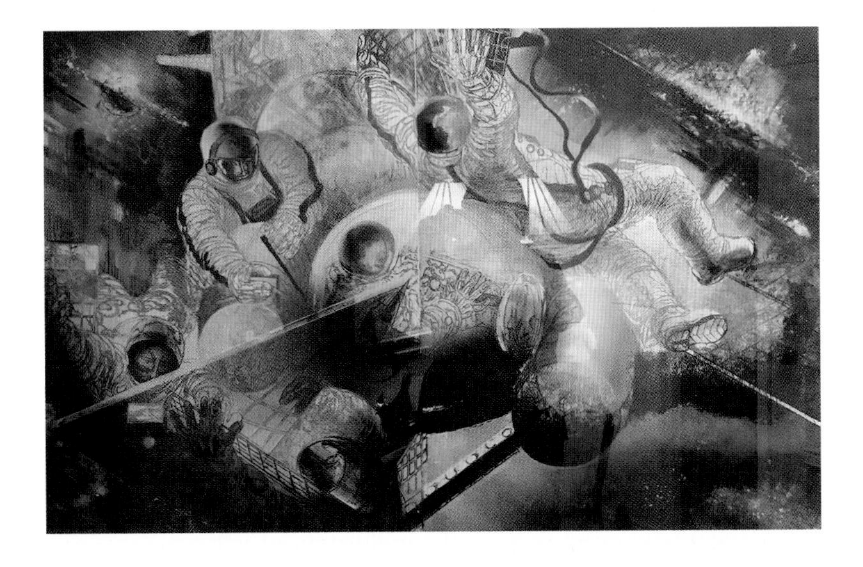

FIGURE 7.1

"Assemblage of Orbital Station," the etching that Anatoly Ivanovich Veselov slipped furtively to the author during the Sputnik anniversary celebrations in Moscow in October 1984. Source: Author's personal collection.

convey with this unusual and entrancing piece of art? Had what I taken as furtiveness perhaps been nothing other than shyness on his part? Everything about our encounter that day and his artistic intentions remains a mystery.[2]

*

Although we were still nearly a year from launch, "reflight training" for the first post-accident shuttle mission began soon after I returned from Moscow. My assignment as Capcom now started to take on real form. All of my flight training credentials had expired after two years of downtime, so I had to start over again back at square one. Like a mountaineer, I would need to climb back up to a basecamp level of proficiency before I would be ready to move on to the peak experience of real-time flight operations. The same was true for the entire mission control team, which had a fair number of new faces to boot because of the turnover that had occurred during the long downtime. Several months of generic simulations lay ahead of us, all designed to get us back up to speed on space shuttle systems and flight procedures and sharpen our problem-solving skills. "Generic" does not mean easy. Each of these sessions was scripted by a group of evil geniuses known as the sim team. The sim team's job was to stress the crew and the entire flight control team by probing knowledge gaps and forcing them to deal with unrecognized interactions between systems or weaknesses in flight procedures. They are very good at their devilry, and fiercely proud about it.

It felt really good to see all the strands of activity that lead to a shuttle flight gaining speed: testing the new software for the shuttle's onboard computers; dusting off the preliminary draft versions of checklists, ground rules, and flight procedures that had been shelved after *Challenger*; and starting the intense analysis and tough training sessions that would transform them into flight-caliber checklists. I felt even better when I learned that the EVA tools for our Hubble deployment mission would be ready for a critical design review in December. The launch date for STS-26 was still iffy, but I welcomed every sign, no matter how faint, that the shuttle program was moving in the right direction and that Hubble might soon stop slipping into the future.

March and April brought more welcome signs of progress. On the shuttle side, the decision was made to adopt parachute bailout rather than ejection seats as the shuttle's emergency escape system. The timeline to the first launch stabilized considerably with this decision out of the way. Down at the Kennedy Space Center, technicians began assembling the solid rocket boosters for the STS-26 flight of *Discovery*, now slated to launch in early September 1988.

The pace of simulations and technical meetings also picked up considerably, now that we were within five months of launch. I would be lead Capcom on the planning shift, supported by fellow astronauts Kathy "KT" Thornton and Pierre Thuot. The title "planning shift" reflected the substance of our work: to replan the mission's activities based on everything that had happened thus far and update the next day's flight plan and checklists accordingly. All these changes, along with news snippets, the latest sports scores, and short family messages, would be transmitted to the shuttle just before crew wakeup time. The clacking sound of the shuttle's teletype machine printing out the invariably huge daily update message was often what woke up the crew in those days.

Our shift would start just as the crew was going to bed and end soon after wakeup time on their next working day. Crew sleep periods rarely aligned with the times that normal earthlings sleep, so we were sure to be working odd hours on all the long-duration simulations ahead and during the mission.

The fun part of my job was choosing the wakeup music. Clever choices that echoed a mission event or played on an inside joke would earn style points from the mission control team and crew, with bonus points available if a selection got picked up on the evening news broadcasts. I felt that more than the usual bit of morning fun was in order for this mission, though. Getting American astronauts back into orbit safely would be a momentous accomplishment and no doubt a cathartic moment for NASA and the country, after the pain and sadness of that awful day in January 1986 when *Challenger* went down. It was important to me that both the wakeup music and my first call to *Discovery* somehow express this. Pierre joined me eagerly on the music search. We soon had a list of decent contenders for the first morning's song, but nothing we were overly excited about. Our prospects brightened considerably when word of our search reached Mike Cahill, a writer and producer at a local radio station who got his space fix by moonlighting as a tour guide at the space center. Mike's daily job required him to create commercials, jingles, and the occasional comedy bit. The glum mood at the space center after the *Challenger* accident inspired him to write some funny space songs to lift everyone's spirits. He came up with fabulously idiomatic lyrics set to the tunes of a variety of popular television and rock melodies, such as "Bandstand Boogie" by Les Elgart, "Green Acres" by Vic Mizzy, and the Beach Boys classics "Fun, Fun, Fun" and "I Get Around." With the help of musician pals Mark Richardson and Patrick Brennan, he recorded them after hours in the radio station's eight-track studio. Mike was only too delighted to let me use his songs for the mission. My musical problem was solved.

What to say after the music stopped on the first morning was a far tougher problem. The routine call—"Good morning, *Discovery.* Houston standing by"—certainly would not do. I wanted to somehow express the jubilation that I was sure we would all feel that morning. Adrian Cronauer's zany, exuberant radio greeting in the movie *Good Morning, Vietnam* came to mind, but plagiarizing Robin Williams and shouting wildly in Mission Control were both out of the question. This left only the possibility that Mr. Williams might share my sense of the occasion and be willing to reprise his role on our behalf. I recalled that one of my Challenger Center colleagues, Rick Hutto, had a lot of contacts in the entertainment industry. Maybe Rick could get my idea conveyed to Williams's agent. Pursuing this plot and chasing musicians would become the comic backdrop of my life over the next five months.

✷

April 1988 also brought progress in the Hubble world. The higher-fidelity training mockups Bruce and I had been pressing for over the past year were finally ready for EVA simulations in the neutral buoyancy facility, along with prototypes of tools and support equipment (such as the toolbox and storage pallets) for the maintenance missions. With better equipment in the water, we could at last begin to nail down how much time it would take to perform each task. This had taken on huge importance over the preceding months for one simple but very worrisome reason: The preliminary reliability assessments indicated that it would take at least four EVAs to accomplish all the tasks that were likely to be slated on a typical maintenance mission. This was twice as much EVA time as the shuttle was able to provide on a single flight back then. Something would have to give if NASA was to deliver on the promise of maintaining Hubble on-orbit. Everyone on the M&R team knew the agency was unlikely to deal with this issue until the telescope was in orbit, but we took it as our responsibility to pass along the best possible timeline estimates to the people who would eventually have to confront this thorny problem, Several rounds of neutral buoyancy simulations were scheduled from late spring onward for this purpose. Bruce and I decided to split the runs between us and include several other astronauts in the work, to ensure the results reflected typical astronaut performance. I had not done a suited test in months, so I was delighted to take one of the first runs, jumping into the tank with Sonny Carter to evaluate solar array maintenance procedures.

Back on dry land, the Hubble engineering teams from Lockheed, Marshall, Kennedy, and Johnson began the suite of meetings that would ensure that the telescope, shuttle, and launch site teams were all on the same page about how the telescope would be tested at the launch site, loaded into the shuttle, and deployed in orbit. A long list of questions remained to be answered: When would the telescope arrive at Kennedy? What cargo facility there could provide the necessary cleanliness levels? What tests did Hubble really need to undergo at Kennedy? How many tests would need to be done once it was installed in the shuttle? What was the highest deployment altitude the shuttle could reach, given its fuel capacity and the telescope's weight? If Hubble's flight batteries weren't ready in time for ground installation (a distinct possibility at that point), could we do an EVA on the deployment mission to install them? How would we deploy the telescope if the shuttle's robotic arm broke down? Every organization that had a stake in the successful deployment of this great observatory had a seat at the table and a voice in these matters. We would spend countless hours in the months leading up to launch analyzing each issue closely and debating, often fiercely, the best solution.

The mission that Bruce and I, along with Steve Hawley as robotic arm operator had been assigned to back in early 1985, STS-61J, no longer existed. The Hubble Space Telescope deployment mission was now known as STS-31. Air Force Colonel Loren Shriver had been named as mission commander and Marine

Corps Colonel Charlie Bolden as mission pilot. Loren convened our first full-crew meeting in April. We reckoned it would be another year before we began training intensively for STS-31, but the time had come for us to meet regularly to keep abreast of issues and ensure that all the strands of work on our mission were coming together as they should. Our most consistent topic of conversation was our launch date or, more accurately, the latest rumors about our launch date. All of them had us well over a year away. Two of the three schedule options in circulation throughout April and May showed the STS-31 launch date as "To Be Determined." We took what comfort we could in not having disappeared from the schedule altogether and soldiered on.

The ever-increasing intensity of preparations for STS-26 and the two flights scheduled soon after it left me little time to mope about my own launch schedule. *Discovery* rolled out to the launch pad in July. As expected, my shifts in mission control commonly ran from suppertime to breakfast time. I would catch a few hours of sleep, spend the next day in technical meetings or a training session, and go back to the control center. I somehow squeezed in my mandatory flying hours and Navy reserve duty along the way. Hurricane Gilbert's direct hit on Houston in August slowed us down for a few days, but added the burden of cleanup and home repairs to the workload for many of us. I was on pins and needles the whole time, hoping we wouldn't discover another technical issue that delayed the flight yet again. Launch seem to be both tantalizingly close and agonizingly far away at the same time.

✶

At 11:37 Eastern Daylight Time on September 29, 1988, the space shuttle *Discovery* lifted off from Pad B, Launch Complex 39 at the Kennedy Space Center. Everyone wearing a NASA badge, and countless more people across the country, held their collective breath as the mission timer clicked up toward eighty-three seconds and *Discovery* accelerated toward Max Q, the time and place where *Challenger* had exploded. It passed that fateful milestone safely and roared on into orbit, letting us all exhale with immense relief. It had taken 975 days, but we were flying again!

After their smooth ascent, the crew set about the task of switching the shuttle's systems over to orbit operations mode and getting ready for the most important payload operation of the mission: deployment of the NASA tracking and data-relay satellite that would replace the one lost with *Challenger*. The TDRS-C satellite, attached to a rocket motor that would boost it up to its perch in geosynchronous orbit (roughly 22,000 miles higher than the shuttle), glided smoothly out of the cargo bay just over six hours after liftoff. After activating the experiments carried in the shuttle's middeck, the crew settled down to eat dinner and go to bed. Back on Earth, my planning shift colleagues and I arrived at mission control just as the crew finished eating, eager for our first real operating shift. Since everything had gone so well thus far, we expected a quiet and relatively easy night,

A panel of programmable digital clocks above the big map on the front wall of the control room kept everyone aware of key mission milestones. The Mission Elapsed Time (MET) clock ticked off the time since liftoff, like a stopwatch. MET was the primary time standard for shuttle flights—our own time zone, if you will. Planning a mission based on MET solved the problem of not knowing precise launch times in advance and cut down on time-zone communication errors among the often far-flung members of the ground team.

Depending on the flight plan, other clocks might mark the countdown to satellite deployment, the elapsed time on a spacewalk, the time remaining before the deorbit burn, or some other important maneuver. The primary clock for the planning shift was the crew sleep timer. If a problem cropped up while the crew was sleeping, we would weigh its severity and the needed speed of response against the time remaining on this clock to decide whether we had to wake them. It also drove our schedule for producing the daily update package, which we would start transmitting to the shuttle when a few minutes remained on the clock. The wakeup music would begin when the timer hit zero.

Except for the countdown clock during my own three spaceflights, I have never watched a clock as closely as I watched the sleep timer during that first shift. To my mild astonishment but great delight, Robin Williams had jumped at the opportunity to help us celebrate the shuttle's return to flight with a rousing crew wakeup call. On top of thinking it was a wildly fun idea, he had a personal connection to the crew: he and Pinky Nelson had been contemporaries at college in Southern California. He sent me a five-inch reel of tape with about a dozen variations of "Gooood morning, *Discovery!*" on it. At least half of them were so edgy that playing them on air-to-ground would surely have gotten me fired, but the others provided more than enough material for my needs. Mr. Williams's agent had also been thrilled by our scheme because of the publicity boost it would offer. He gave me a scare when he laid out his plan for promoting Robin's participation ahead of the launch. I told him in no uncertain terms that this was to be a surprise, and that I would destroy the tape if there was so much as a peep from their end beforehand (I had the foresight to demand they give me the studio master tape, so could make good on my threat). For the first morning's wakeup, Pierre and I made an audio tape with two of Robin's craziest opening shouts and a corny number Mike Cahill had written to the jaunty tune of the *Green Acres* sitcom title song. We had managed to keep our escapade secret the entire time it was afoot. Only Pierre, KT, and I knew what was about to hit the airwaves.

The sleep clock finally hit zero. The tone that signaled an active link to the shuttle beeped into everyone's headset a fraction of a second later, followed instantly by Robin Williams bellowing "*GOOOOOOOD MORNING, DISCOVERY!*" Quizzical looks and then huge grins spread across the faces in the control room at the sound of his voice, so very familiar but utterly unexpected within the hallowed walls of space shuttle mission control. On he went: "Good morning, *Discovery*. Rise and shine, boys. Time to start doin' that shuttle shuffle, you know what I mean? Hey, here's a little song comin' from the billions of us to the five of you. Rick, start 'em off, baby. The Hawkster, to you." Then the "Green Acres" melody started, and Cahill and his pals launched into lyrics that seemed absolutely perfect for the shuttle team's happiest day in nearly three years:

On orbit is the place to be,
Free-wheeling on *Discovery*.
Earth rolling by so far below.
Just give her the gas and look at this baby go.
We can't believe we made it here,
So high above the atmosphere.
We just adore the scenery.
Yeah, Houston's great but give me that zero-gee.
Hey, look out the window!
That's neat!
Cap'n, I'm hungry.
Let's eat!
Maybe we'll land at ... White Sands?
Uh-uh.
Look ma, no hands!

This is the life! Oh, what a flight!
Earth orbit, we are here![3]

Rick Hauck jumped onto the air-to-ground link as soon as the tune ended, with an exuberant, "*Gooooood morning, Houston!*" The STS-26 crew was clearly awake and in very good spirits.

After a smooth mission and reentry, *Discovery* landed safely in California on October 3. "A great ending to a new beginning," the words of Capcom Blaine Hammond when the shuttle finally came to a stop, seemed spot on. It was my thirty-sixth birthday, and I couldn't think of any better presents than a flight crew safely home and the shuttle program back in business.

The space shuttle's return to flight would be featured prominently in every national television network's year-end retrospective on the major events of 1988, and every one of these would use Robin Williams's exuberant greeting and Mike Cahill's amusing tunes to capture what it had meant to NASA and the nation. I got at least triple-bonus style points as the planning shift capcom. Astronauts being competitive types, I'm sure my successors on the planning shift console would have tried to best me, if they had only had the chance. Sadly for them, objections from the Department of Defense and the bevy of musicians lobbying to also write custom wakeup songs prompted NASA headquarters to ban unpublished music after STS-27.

✷

The tide of work began to turn back toward Hubble for me as 1988 came to a close. My stint as Capcom was due to end in March, with the flight of STS-29. It had been a challenging and fun assignment, but I was eager to dive fully back into Hubble work. Hallway rumors had finally turned into an official launch date of December 1989. If that date held, our training schedule would soon pick up speed. Betting on a positive outcome, we started to work on one of our top crew priorities: the design of our mission patch.

Bruce was keen to lead this effort, and the rest of us were only too happy to grant him this wish. It was obvious that the main elements in our patch should be the shuttle, the telescope, and some celestial objects, but less obvious how to link them together into a reasonably compelling design that would still be recognizable when stitched into a patch about three inches in diameter. We wanted our patch to convey a sense of motion and also to allude to the science that Hubble would do. We thought that a swoosh—a tapering ribbon of color—progressing from white to blue to red, could perhaps do both. The swoosh would suggest the shuttle was flying through the scene, and the color progression would signify the redshift, a property of light that astronomers use to determine how fast the universe is expanding. One of the telescope's key scientific goals was to determine this expansion rate (known as the Hubble Constant, in honor of the telescope's namesake Edwin P. Hubble) to an accuracy of 10 percent. Bruce duly took our ideas to the artist we had chosen, who soon produced a design we all liked (figure 7.2).

The dipped wing and tapered swoosh worked well, making it look like the shuttle was flying past the telescope and toward you, the viewer. The vivid colors in the swoosh framed the telescope nicely and made the shuttle stand out clearly. Astronauts tend to be perfectionists, however, so of course we spent quite a bit of time fretting over the fact that the swoosh right behind the shuttle was red rather than blue. The patch clearly looked better with the colors sequenced this way, but in reality, an object moving toward the viewer would shift the light toward the *blue* end of the spectrum. Would we be accused of not knowing our basic physics? Would the

FIGURE 7.2

The STS-31 mission emblem. Source: NASA.

astronomers be offended? Was this really an important question? We let artistry win the day.

✷

The pace and intensity of mission planning picked up dramatically with the announcement of a firm launch date. All the activities involved in preparing Hubble and the space shuttle *Discovery* for flight had to dovetail smoothly at specific points in time, like the date the telescope would ship from California or be installed in the shuttle's cargo bay. The reams of design documentation and draft procedures that had been produced over the years now had to be turned into the kind of highly detailed operating plans needed for a successful spaceflight. Dozens of significant technical questions had yet to be answered, not least among them the altitude at which we would deploy the telescope and whether there were any circumstances under which we would bring it back to Earth with us. The people side of the equation was just as complex as the technical side: shuttle flight controllers at the Johnson Space Center in Houston, Hubble flight controllers at the Goddard Space Flight Center in Maryland, and engineers from both Lockheed and the Marshall Space Flight Center all had to learn to understand each other and develop the skills needed to work together at 17,500 miles per hour (28,163 km/hour).

We would tackle these challenges with two proven technologies: conversation and simulation. Mission simulations with the full shuttle plus Hubble flight control teams would not start for another six months or so, but the conversations (a.k.a. meetings) began right away. Much of our time in the months leading up to launch would be spent in "Flight Techniques" or "Operations Working Group" meetings designed to surface questions or potential problems, identify key interdependencies, prompt technical analysis and debate, and force decisions about how we would conduct the flight. I thought of this process as the reverse of peeling an onion: we started with the basic elements of the operations and then added on layer after layer of ever more refined detail. The intensive discussions and debates along the way gave us a deep understanding not only of every technical system and operation, but also of each other's competencies and personalities. These would stand us all in good stead, both when our simulator training started and during the real flight.

December brought a three-day round of maintenance mission EVA tests in the Neutral Buoyancy Simulator at Marshall. This round was memorable for two reasons. The first was that this would be the last chance to work on maintenance mission EVA tasks until after the deployment mission. The Hubble training mockup would soon be shipped to the sister neutral buoyancy facility at the Johnson Space Center in Houston, known as the Weightless Environment Training Facility or WETF, where the remainder of our training would take place (figure 7.3). One driver for this change was the need to reclaim the time Bruce and I spent traveling to Marshall, so we could apply it to other training priorities. The more important driver was the need to electronically connect the neutral buoyancy tank and shuttle simulators together and tie them both into mission control, so that we could practice deployment mission operations realistically with the entire Hubble team. Until this point, our EVA training sessions had involved only Bruce and me, the EVA experts from Houston who would staff the EVA post in mission control during the flight (Jim Thornton, Robert Trevino, and Sue Boyd, who had replaced Kitty Havens when she moved into a management role), and the Lockheed team of Ron Sheffield, Brian Woodworth, and Peter Leung. The eight of us had spent countless hours in water tank simulations and working on the telescope itself since 1985 and were perfectly ready to do our bit. Now, however, our bit had to fit seamlessly into a much larger and very dynamic jigsaw puzzle, comprising the space shuttle, the real and very complex Hubble Space Telescope (not the inert version we had worked on in the tank), and the scores of people in shuttle mission control and the Space Telescope Operations Control Center. Tying all the simulators and control centers together electronically was essential to forging these varied groups into a unified flight operations team, and only Johnson's WETF had the circuits needed to do this.

A McCandless tantrum was the second memorable thing about this final visit to Marshall. No self-indulgent act, this was a calculated and well-executed bit of theater meant to avert another bureaucratic tussle between Marshall and Goddard, one that we feared could jeopardize the success of future maintenance missions.

FIGURE 7.3

Scuba divers monitor Kathy Sullivan (spacesuited figure at right) and Bruce McCandless (top center) as they practice EVA tasks on the Hubble mockup in the Johnson Space Center Weightless Environment Test Facility (WETF) during training for STS-31. Source: NASA.

The trigger was Bruce's realization, during our pre-dive briefing, that the lone Goddard person present was a fairly junior engineer. NASA had decreed that Goddard would lead Hubble maintenance operations once the telescope was in orbit, even though that center had played virtually no role in the maintenance preparations to date. It had been Marshall, as NASA's lead center for Hubble, that ensured that the telescope was designed for maintainability. Then, starting in 1985, it had been Marshall engineers, along with the Johnson Space Center and Lockheed members of our M&R team, who worked to design, produce, and test the tools and procedures needed to ensure that Hubble maintenance was a practical reality, not just a design goal. In Bruce's view, Goddard should have seen the maintenance mission simulations we were about to run as ideal opportunities to come up to speed on this early work and get acquainted with the proven hardware and procedures that they would inherit as the foundation for future maintenance missions. He was appalled that they had not sent a higher-level manager who had some authority over their preparations for this new role and could ensure a smooth transition of the equipment and accumulated knowledge. I think his real worry was that this very substantial and costly body of work would fall victim to the "not invented here" syndrome, Goddard's notorious penchant for ignoring the work of others and designing (or redesigning) everything in house.[4]

If Bruce had been worrying about this beforehand, he had not said a word to me. I pieced the picture together as he lambasted Goddard for failing to give Hubble the priority attention it deserved, Marshall for acquiescing to this without a fight, and NASA headquarters for failing to make their children behave better. Bruce was on a tear. Nobody in the room dared to interrupt him. Some people stared at their shoes as he railed on, others fiddled with their notebooks. Everybody knew he was right, but nobody in that room could do anything about the issue he was raising. I waited until he had hammered his points well and fully home, and then chose a moment to intervene. "You've raised a really critical issue," I said, "but not one that can be solved by the people present here. So, I think the real question at hand is whether we are going to get into the tank and do the test today, or not." The room held its collective breath as we waited to learn if he would choose to cancel or proceed, and cleared out in a nanosecond when he said, "Let's do it."

I don't know who called whom, but the phone lines between Marshall and Goddard must have been burning up while we were busy working underwater. By the time we returned to the conference room some four hours later for the post-test debrief, a senior engineer from the satellite servicing shop at Goddard was on an airplane bound for Marshall to observe the remaining two tests. Bruce's tirade had worked just as he intended.

*

The telescope's solar arrays, being built in Bristol, England, by British Aerospace, would be the last major component to be installed on Hubble. These were scheduled to undergo final factory testing in January 1989, before being shipped to the Lockheed plant in Sunnyvale. Bruce and I had used an earlier round of tests back in 1985 to test prototypes of our EVA tools and study the hinges, latches, and other mechanisms involved in deploying the booms and unfurling the solar cell blankets. This final preshipment test would be the only opportunity for Loren, Charlie, and Steve to get a firsthand look at these critical mechanisms. Since any problem with one of the solar arrays during the deployment would trigger an EVA, Loren decided the entire crew should make the trip to England to see them up close.

You almost had to see the telescope's solar arrays to believe them. While most satellites use rigid solar array panels that unfold accordion-style after launch, Hubble's resembled roll-up window shades. A combination of factors, including the ambitious science goals, the Earth-return premise of the original maintenance plan, and the design of the shuttle itself forced Hubble's designers to take this unique approach. The arrays had to be quite large to meet the telescope's four-kilowatt power demand with the solar cell technology then available. The assumption in the original maintenance plan that the telescope would be brought back to Earth every five years for major maintenance meant that the arrays needed to be retractable, so the telescope could fit back into the shuttle. The tiny bit of cargo bay volume left vacant by the telescope argued against rigid panels, and also set very tight size and shape limits on the design.

The solar arrays that Hubble would begin its life with consisted of flexible solar cells bonded to blankets, just 1/32 of an inch thick (or 0.76 mm, thinner than a page in a typical hardcover book), of a gold cellophane-like material called Kapton.[5] Each solar array wing was ten feet wide and twenty feet long. Two of these would be coiled up on a pair of rollers mounted on each of the telescope's twin solar array booms, for a total of four wings. At launch, the curtains would be rolled up and the boom folded along the side of the telescope. In orbit, the arrays would deploy in two stages. First the booms would pivot out to a position perpendicular to the body of the telescope. The similarity to a household roller curtain would end at this point, since obviously nobody would be waiting outside the shuttle to pull the arrays off the rollers. Instead, a "spreader bar" that ran along the outer edge of each curtain would be pushed out from the boom. This was accomplished by a pair of telescoping metal tubes, one attached to either end of the bar, that extruded from remarkably small storage canisters, like spaghetti coming out of a pasta maker.

When I first saw the arrays back in 1985, I found it hard to conceive how a pair of twenty-foot-long metal poles could possibly fit inside each of the two small canisters that were mounted on the solar array booms. Their constant diameter was evidence that these tubes (called "bi-stems") were not made like a collapsible walking stick, with segments of differing size along the length. A closer look revealed each pole was made of two thin, C-shaped metal strips nested together lengthwise. These reminded me of a tape measure, which turned out to be a good hint at how the Hubble design works. The metal strips were rolled up onto storage spools inside the canister. As each strip came off its spool, it passed over a roller that popped it back into the C shape. The rollers were placed so that one strip popped into shape slightly before the other, allowing them to wrap around each other to form a cylinder. Like a metal tape measure or a piece of paper rolled into a tight tube, the bi-stems would kink easily if pushed from the side but were quite strong lengthwise—clearly strong enough to overcome the spring tension on the rollers and push out the full twenty-foot length of a solar array wing. The bi-stems were also too flimsy to support the weight of the solar array blankets on Earth, so the only way to test the deployment mechanism on the ground was on a water table, with blocks of Styrofoam placed every few feet to prevent the blankets from getting wet as the arrays unfurled from the storage rollers (figure 7.4).

This clever design solved the original challenges, but also introduced a major problem that would lie undiscovered until Hubble was in orbit. A spacecraft orbiting at Hubble's altitude passes from searing sunlight to the frigid darkness of Earth's shadow and back roughly every forty-five minutes. The bi-stem tubes would heat and cool hundreds of degrees in a matter of seconds at each of these transitions, sending vibrations through the telescope that made stable astronomical observations impossible for ten to

FIGURE 7.4

The Hubble solar arrays being unfurled on the water table in the British Aerospace test facility outside of Bristol, England. Source: European Space Agency, http://www.esa.int/Our_Activities/Space_Engineering_Technology/How_Hubble_got_its_wings.

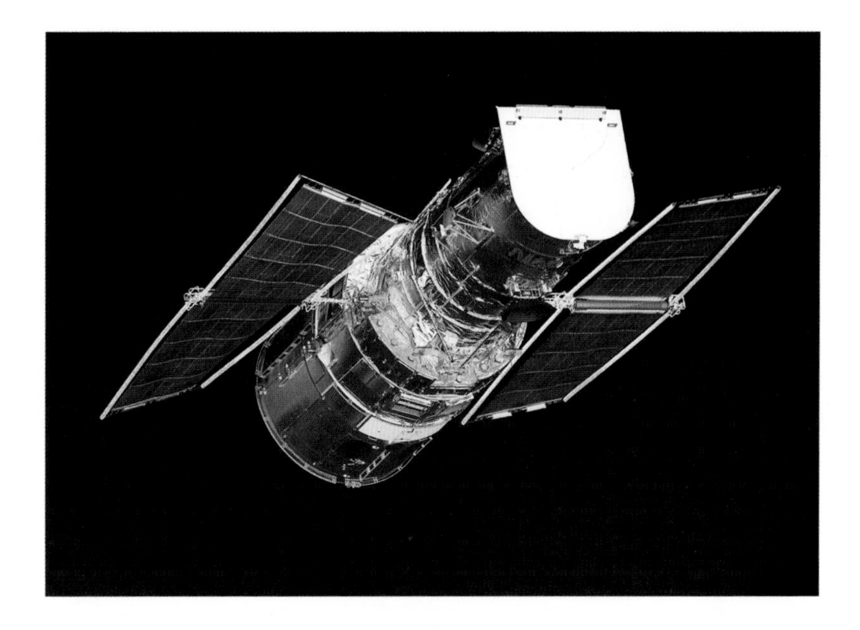

FIGURE 7.5

Hubble Space Telescope in orbit. The bi-stems are the long silver rods running along each edge of the gold solar array blankets. The thermal warping is most apparent in solar array on the left. Source: NASA.

fifteen minutes and sometimes tripped the telescope's control system into emergency mode (figure 7.5). The control problems and loss of observing time would prove so severe that astronauts on the second servicing mission in 1997 would replace the original wings with accordion-style, flat-panel arrays.

✷

In April 1989 we reached a milestone that every crew looks forward to, one that signifies the formal start of flight-specific training: we were "going downstairs." This meant the time had come to turn full control of our calendars over to Tom McClure and Erlinda Stevenson, the duo one floor below us who scheduled all the training sessions for every crew that was within one year of flight. Every Friday from then until launch, they would give each of us a calendar for the following week that optimized the use of our individual and collective time and accomplished as much training as possible. Creating these was like building a multidimensional jigsaw puzzle. Some training sessions could be done at any time; others had to be accomplished in a specific sequence. Some could be done solo; others needed only the pair of EVA crewmembers or the trio (commander and pilot plus one mission specialist as flight engineer) who were the core flight deck crew for ascent and reentry. Sometimes we could claim time in the primary mission simulator; sometimes one of the crews in line ahead of us would bump us out. Naturally, Murphy's Law jumbled all of this on a daily, sometimes hourly basis. We were sixth in line to fly when we went downstairs, which meant Tom and Erlinda were working six different jigsaw puzzles and producing weekly calendars for over thirty people. There were no handy scheduling apps back then to help them with this monumental challenge of continuous optimization. They overcame it with pencil, paper, and telephone, plus unflagging persistence and a seemingly inexhaustible supply of good humor. We were in good hands with those two.

I had been looking forward to April for another reason as well: I would get to do an earthbound EVA to test real Hubble flight hardware inside Chamber B, one of the large vacuum chambers at the Johnson Space Center. With its battery of powerful air pumps and liquid nitrogen cooling system, Chamber B could simulate the extreme vacuum and cold temperature conditions of low-Earth orbit or deep space.

Adding to my excitement was the fact that I would get to wear a "Class 1" spacesuit again—one designated solely for use in orbit.

People are surprised to learn that astronauts preparing for spacewalks do virtually none of their training in the Class 1 suit they will wear in orbit and spend almost no time training in a vacuum. Astronaut Mike Massimino, a two-time Hubble repairman, has likened this to training for the World Series without ever setting foot in a real baseball stadium.[6] We did the lion's share of our EVA training underwater, wearing "Class 3" spacesuits that are dedicated to this purpose. Frequent and heavy use makes these training suits much more pliable and easier to move in than a Class 1 suit. Many an astronaut who aced every task in the water tank has found the job much harder and more tiring in orbit because of the greater stiffness. That's why spacewalkers leap at any chance to work in their Class 1 suit in a vacuum.

On top of all that, the chamber test would cap off all the years we had spent prototyping and redesigning hardware. We would finally be working with real, space-ready Hubble maintenance equipment. One particular piece of equipment was also the reason why Bruce and I would get to go into the chamber to do these tests. Vacuum chamber tests typically use machinery to twist, turn, or otherwise exercise the flight hardware in the same way it will be used in space. The Hubble adjustable foot restraint was a complex, one-of-a-kind device, however. No test machinery existed that could operate its pedals and hinges, so this would have to be done manually. Bruce and I considered this a very lucky break.

The primary aim of the chamber tests was not to confirm that our tools and equipment would work in the absence of air (i.e., the vacuum of space), but rather to make sure they would work at extremely cold temperatures. We knew our wrenches fit all the bolts on Hubble when both the telescope and wrench were at room temperature in the VATA clean room at Lockheed. Would they still fit at −130° Fahrenheit (−90°C), or would they have shrunk too much? What about the mechanisms inside the foot restraint platform that allowed it to tilt and pivot—would they still work smoothly at such frigid temperatures? Would the battery in the power tool last long enough in the cold to crank both solar arrays out fully? The only way to find out was to test them in these very cold conditions, and the vacuum chamber was the only way to create those conditions here on Earth.

Our test setup looked puny inside the large volume of Chamber B. A metal ramp led from the airlock door down to our working area on the chamber floor a few feet below. The foot restraint was suspended from a metal frame, positioned carefully so that we could move all the pedals and joints without either hefting its full weight or worrying about knocking it down. Highly accurate mockups of a variety of Hubble electrical connectors and fasteners were arrayed on an adjacent workbench. The test layout was designed so that we could pivot from one task to the other with few, if any, walking steps. This was a matter of safety, not laziness. The life support backpack on our suits would be suspended from an overhead monorail that would literally take some of the backpack's three hundred pounds of weight off our backs. The monorail made walking possible, though still very awkward and difficult. Bruce and I would each perform the same set of tests three times: once in a shirt-sleeved dry run, again in a fully suited dry run with the chamber at sea level pressure, and, finally, fully suited with the chamber pumped down to vacuum.

I was up first for the vacuum run, thanks to a hiccup with the monorail system that scrubbed Bruce's run. My suit technicians and a flight surgeon were ready and waiting when I arrived at the chamber at 7:30 in the morning. After the medic confirmed that I had not gotten sick since the day before, it was time to suit up. Suiting up always has a serious, almost ritualistic feel to it. Step by step, you are sealing yourself into your very own body-shaped, single-person spaceship. I occasionally practiced doing it with my eyes closed and without any help from my suit techs, just to hone my skills and make sure I could do it alone if need be. In reality, you hope to never do it alone. You *want* your suit techs or crewmates to check and double check that all the connections are made and the seals are tight. My always-alert suit techs, whom I trusted completely, were hyper-vigilant that day. It's one thing to suit me up and lower me into the water tank, where I am surrounded by scuba divers, two of whom would do nothing but watch me for any sign of a suit leak. It is quite another to leave me alone in a room and then pump all the air out of it, and that is what we were about to do.

Four hours of mind-numbing boredom came first.

The pressure inside my spacesuit would be a mere 4.3 pounds per square inch (296 millibars), roughly one-third the pressure level surrounding me at sea level in Houston. That's equivalent to an altitude of 30,330 feet (9,245 m), which is 1,300 feet (396 m) higher than the summit of Everest. In addition, the "air" inside the suit would actually be pure oxygen instead of the 80 percent nitrogen, 20 percent oxygen mixture of our atmosphere. My blood and the tissues in my body cannot hold as much nitrogen in that low-pressure, high-oxygen environment as they hold naturally at sea level. As we depressurized the airlock, the suddenly excess nitrogen in my body would escape, forming bubbles in my bloodstream just like carbon dioxide bubbles form when you open a bottle of seltzer water. Too rapid a change from the sea-level pressure of Houston to the low-pressure, zero-nitrogen suit environment could give me a serious or even fatal case of the bends.

We had two protocols for managing this risk in the shuttle program. In flight, we used a base camp approach, lowering the pressure in the entire shuttle cabin to 10.2 psi (equivalent to an elevation of 9,700 feet, or 2,957 m) for at least twenty-four hours before a spacewalk. The nitrogen level in everybody's bloodstream would slowly equilibrate to this lower pressure. The EVA crew members would accelerate the washout of nitrogen in their bodies by spending an hour on oxygen masks before the decrease to 10.2 psi began and then, on EVA day, would spend another forty minutes breathing pure oxygen in their spacesuits before depressurizing the airlock to vacuum.

There were two reasons this protocol would not work for the chamber runs: we couldn't spare the twenty-four hours and there was no small compartment or room that we could depressurize partially to serve as an overnight base camp. That forced us to use the alternate protocol: a four-hour in-suit prebreathe. I donned my suit, checked everything one more time with my suit techs, and settled myself onto a small bench inside the airlock to while away the four hours. My options for passing the time were few: swap jokes with the test engineers, listen to the music they piped in (country music plus tinny communications circuit is a very bad combination, especially if you're not partial to country music in the first place), meditate, or nap. I relied heavily on the latter two.

The test conductor called everyone back to full attention as the four hours drew to a close. Ensuring the chamber systems were all working properly and the airlock was ready for depressurization was the first order of business. Next came suit checks. I toggled the switch that brought up all the key operating parameters on the computer chest pack mounted to the front of my suit. Everything was GO. The chamber crew began pumping the air out of the airlock where I sat. Each time I did a chamber run, I marveled at the fact that the room looked exactly the same with and without air in it. My eyes could not tell me whether this was a space in which I could live healthily or one in which I would die quickly. The chamber operating crew would sometimes drive this point home in training sessions by putting a shallow cake pan filled with room-temperature water on the floor before closing the hatch. As the pressure dropped toward vacuum, the water would start to bubble, then boil violently, and finally—and instantaneously—flash-freeze to ice and fall into the pan. The test conductor would then dryly remind the test subject, "That's what would happen to the fluids in your body if you weren't in that suit." Message received: stay on your toes and take very good care of your suit; without it, you die.

The airlock finally reached vacuum, allowing the test operator to open the door into the chamber. I lumbered awkwardly toward the door, hoping the monorail system would work properly today; it did, thank heavens. I stepped carefully over the low doorsill and made my way slowly down the ramp toward our test equipment. I was keenly aware that a fall would force us to break vacuum and repressurize the gigantic chamber back to sea level so the suit crew could get me upright again. Besides being a huge embarrassment, this would waste all of the day's effort and cost the chamber crew at least another full day's work to get everything ready again. I did not want that on my head.

I pity anybody who had to watch me work over the next six hours. There is nothing exciting about watching someone—even a woman in a spacesuit working in a room without air—repeatedly disconnect and reconnect electrical connectors (fake ones, at that), loosen and tighten bolts with wrenches and a power tool, and twist around a footstool-style contraption that looks like Rube Goldberg himself must have made it. It wasn't even exciting, in an adrenaline-rush kind of way, for me, and I was the woman in the spacesuit. It was, however, all-absorbing. Each part of the test had two objectives: check that the tools worked properly and elicit "user's notes" feedback to help train future astronaut and mission control crews. I concentrated intensely on executing each test step carefully and describing every detail about tool operations over the communications circuit, while simultaneously trying to mentally record every suit sound and sensation. The chamber support equipment competed for my attention as well. The cable and spring connecting my backpack to the monorail overhead sagged as I worked, pushing me into a forward hunch and forcing me to stop every thirty minutes or so to straighten my back and force the weight-relief gear back into position. Six hours of that left me with three days of the worst back cramps I have ever had.

The cold began to dominate my attention as the test wore on. Sunlight and movement would help me keep warm on a real spacewalk, but there was none of the former and too little of the latter in the chamber. Standing by the test equipment and rarely moving more than my arms, my body was simply not producing enough heat to keep the frigid temperature in the chamber at bay. Ignoring my cold feet and hands was not too hard at first, but my jaw began to tremble uncontrollably as the chill set in more deeply. That put an end to the running commentary I had been providing as I went about each task. I now had to stop and concentrate fully on controlling my chattering teeth in order to talk. Fortunately for me, the test was nearly over by that point, and I was soon warming up back in the airlock. We couldn't have known it then, but my case of the chills was a harbinger of things to come. The same combination of deep cold, no sunlight, and low workload would lead to a case of mild frostbite on Story Musgrave's hands during a chamber run leading up to the first Hubble maintenance mission in 1993. Story recovered, but the incident would prompt NASA to add insulation to the spacesuit gloves and adjust the shuttle's orientation in flight to keep the payload bay environment warmer during spacewalks.[7]

✷

At what altitude should we deploy the Hubble Space Telescope? For the astronomers and Hubble program managers, the easy answer was "as high as possible." They wanted to minimize both atmospheric drag on the school-bus-sized telescope and the gravitational forces that the gyroscopes would have to work against.[8] Fair enough, but that just raised another question: "How high is possible?" It was this question that forced the Hubble and shuttle teams to wrestle with the competing technical needs and capabilities of those two very different spacecraft, the various operating challenges posed by the deployment and maintenance missions, and the physics of the sun. The five of us on the deployment mission flight crew also cared deeply about this question, but for a very different reason: if we used too much shuttle fuel taking Hubble to a really high deployment altitude, we would not have enough left to get back to Earth. And as much as we were looking forward to being in space again, we cared even more about getting back home at the end of our mission. So, we paid very close attention to the debates on this topic.

Beyond my vested interest in survival, I was drawn into the issue because it was a particularly fascinating example of the complex scientific and engineering problems involved in planning any spaceflight, and what it takes to solve them. Planning a spaceflight is rather like solving a jigsaw puzzle without a picture on the box to guide you. You don't know in advance exactly what the end result is supposed to look like. You are fitting the pieces and creating the picture at the same time. The major Hubble pieces in this jigsaw puzzle were the weight and size of the telescope, its drag coefficient in orbit with the solar arrays and antennas deployed, and the capacity of its gyroscopes to hold it steady despite the drag and gravitational forces. The shuttle pieces were the "upmass" (the weight of the cargo at launch), the total propellant load onboard, the amount of propellant needed for the time in orbit, and—our favorite point—the amount needed for the deorbit burn that would bring us back to Earth.

Two additional elements—the sun and the calendar—are what made this mission planning puzzle more challenging than any I had worked on before. Our Sun goes through regular, roughly eleven-year cycles of higher and lower activity, marked by more sunspots, solar flares, and other bursts of solar energy. Because our launch date had slipped from 1986 to 1990, Hubble would now launch near the peak of one of these cycles, the solar maximum of Cycle 22. The more intense solar activity would increase the atmospheric density at the orbital altitudes under consideration, possibly by a factor of fifteen.[9] Higher density would make the telescope harder to control and would cause the orbit to decay much faster. The astronomers' argument for a high deployment altitude was driven largely by the fear that the orbit might decay so much before the first planned maintenance mission that Hubble's gyros would be overwhelmed by drag and gravitational forces. This could halt the scientific work and perhaps even jeopardize the telescope itself. Thinking ahead to a maintenance mission roughly three years after deployment raised a competing concern: that the telescope's orbit might not decay enough for a maintenance flight to reach it. We were devoting every possible drop of fuel to gaining altitude on the deployment mission. A maintenance mission would have about the same cargo and fuel load as us, but would have to devote some of its fuel to the engine burns required to rendezvous with and grapple Hubble. The telescope's orbit would have to decay enough to free up this rendezvous fuel.

So, the answer to the question "How high is possible?" was "That depends on the solar cycle and how long it will be until the first maintenance mission." The challenge of analyzing all the pieces of this jigsaw puzzle was tackled by the combined Hubble team in technical meetings known as "Flight Techniques" that were held on a monthly basis starting in late 1988. Led by one of the Johnson Space Center shuttle flight directors, these sessions brought together shuttle flight controllers from mission control in Houston, Hubble operators from the Space Telescope Operations Control Center at Goddard, and telescope project engineers from Marshall and Lockheed to thrash out the engineering and policy issues that needed to be solved before flight. Month after month, these sessions reviewed the most recent forecast for the timing and intensity of the Cycle 22 maximum, the rate of orbital decay observed on current NASA satellites, the latest analyses of shuttle fuel usage, and updated estimates for maintenance mission cargo weights. The often-heated discussions challenged each piece of data and every analytical method, probed the uncertainties in the calculations and estimates, and debated the risks to telescope, shuttle, and crew. Eleven months of debate and analysis slowly narrowed the range of possibilities and brought the Hubble program managers up against a hard decision deadline. In July 1989, they officially requested a deploy altitude of 330 nautical miles (611 km), roughly twice as high as the shuttle usually flew. We were really excited about having more spectacular views of Earth than ever before, but very much sobered by the fuel calculations: our fuel tanks would be a little over half empty a mere forty-two minutes into our five-day mission. That meant the slightest sign of a leak would trigger a rapid emergency response aboard the shuttle. In the worst case of a leak on deploy day, this could have a particularly devastating impact on Hubble. The Hubble flight control team would face precisely this worst-case scenario just seven months later. Lucky for them, it would only be a simulated leak in a training exercise.

*

October 24, 1989: launch minus six months. At long last, we were counting down to launch in months instead of years, a very good feeling indeed. From this point forward, we would devote one hundred percent of our time and energy to preparing for STS-31. Eager to pick up the pace and intensity of our preflight training, we happily handed off all our nonflight duties to other astronauts.

The dozens of issues and strands of planning activity we had tracked separately for years were now coming together rapidly into a solid flight plan. The list of EVA tools we would carry for contingency maintenance on the Hubble deploy mission had been finalized. The particular way we wanted them loaded into caddies had been defined, and the storage location for each item aboard the shuttle specified. Now we also had a list of so-called minor payloads we needed to learn how to operate. These were small experiments that fit inside the shuttle middeck and used up the tiny bit of maximum liftoff weight not devoted to Hubble. We would spend most of the next three months in what were called "standalone training" sessions. Some of these would be hands-on practice with our middeck experiments, aimed at deepening our knowledge of the hardware and the underlying science. Others would be simulator sessions designed to fuse our knowledge of individual shuttle systems into a deep understanding of the vehicle as a whole and sharpen our operating skills as a crew. "Integrated Sims"—simulator sessions that put us and our mission control team through our paces together—would begin around January. A series of three Joint Integrated Sims in March and April would provide the most intense, realistic, and crucial training for STS-31. Each of these would force all the many components of the Hubble deployment team—our crew aboard *Discovery*, shuttle mission control in Houston, Hubble flight control at Goddard, plus the Marshall and Lockheed engineering and maintenance teams—to live through a twelve- to eighteen-hour slice of the mission together.

These Joint Integrated Sims were scripted carefully by the shuttle mission training cadre in Houston. Led by a Sim Supe (for Simulation Supervisor), the job of this team of devilish engineers was to create training scenarios that stressed the combined flight and ground teams, with the goal of revealing each other's strengths and weaknesses and forcing us to develop strong problem-solving skills and communications discipline. They did this by poring over all the flight procedures and operating ground rules to find weaknesses—undetected interdependencies, overlooked issues, half-baked solutions, and the like—and then building scenarios that poked hard at these soft spots. The sim scripts they developed were guarded like state secrets to preserve the surprise factor that would force both shuttle and ground teams to hone their real-time problem-solving skills. We expected that the first Joint Integrated Simulation would be a "nominal deploy" to let the combined Hubble team experience deployment the way everybody hoped it would unfold, with just a handful of fairly minor problems thrown in to keep everyone on their toes. The next obvious scenario to simulate was a time-sensitive contingency EVA, triggered by the failure of one of the telescope's solar arrays to deploy. We, the flight crew, argued strongly that the third Joint Integrated Simulation should be the Hubble nightmare scenario: a propellant leak partway through deployment operations.

Our argument was simple: despite months of urging by many parties, the Hubble team had staunchly refused to address what they would do if we had to release the telescope under less-than-ideal conditions. They had a very long list of release constraints—full sun on the solar arrays, solid communications through multiple antennas, and at least a specified amount of daylight remaining before orbital sunset, to name just three—that they insisted all had to be met before we set the telescope free, and had no idea how to proceed if they did not get everything they wanted. We, on the other hand, knew exactly what each member of our flight crew would do at the first indication of a propellant leak:

fan out and start emergency deorbit preparations. Charlie would immediately start troubleshooting to confirm whether we had a real leak or a bad sensor. Not waiting for an answer, Loren would start programming the computer for a retrofire burn to lower our orbit or take us home, depending on how bad the leak proved to be. Steve would get everything ready to release the telescope and close the payload bay doors. Bruce and I would start stowing the cabin for reentry and landing. Knowing that time would be of the essence with a real leak, we had practiced this many times and pared the time from leak alarm to deorbit burn down to less than twenty minutes. The choice of either making everything perfect for Hubble or heading for home before our propellant leaked away was a no-brainer for us: head home ASAP. Whether we would get all the way home or not would depend on where the leak was in our propellant system. Some leaks could be isolated without losing all the propellant, others not. In the former case, we would lower our orbit and wait until we were in range of a proper runway. In the latter case, we would execute an emergency deorbit immediately and, most likely, bail out of the shuttle wherever we came down. The odds of this really happening in orbit were certainly slim, but the Hubble team's refusal to even think about the eventuality was irresponsible. Our third joint simulation would force them to deal with it openly.

The final twelve-hour joint simulation of Hubble deployment operations started at 8:00 a.m. The sim team's propellant leak cropped up around forty-five minutes later, when the telescope was poised above the cargo bay on the end of the robotic arm. The five of us in the shuttle simulator jumped into emergency deorbit mode and had the shuttle back on the ground in about an hour. We were done with the simulation around 10:00 a.m. The Hubble control team labored on for another ten hours, struggling to operate the telescope in those very unexpected circumstances and no doubt praying that they would find a way to salvage the mission. The next day's debriefing session had to be equally painful for them, as the lead flight director in Houston, Bill Reeves, made them walk the combined Hubble/shuttle team through each event and account for their hits and misses. Humbling though it no doubt was, this Joint Integrated Simulation did exactly what it was designed to do. The draconian scenario revealed some significant problems in the Hubble control center's software and weaknesses in the control team's operating procedures. Building these hard-won lessons into their flight procedures would pay big dividends once Hubble was in orbit.

*

The five of us on the STS-31 crew flew to Florida on Sunday, March 18, 1990, for our countdown dress rehearsal. Just like in the theater, the goal of this dress rehearsal was to bring the entire cast and crew together on stage (or launch pad, in our case) to make sure everyone knew their parts and could perform the show (countdown) flawlessly. The full launch day crews would be at their stations in each of the three control centers involved in STS-31: the Kennedy Launch Control Center (LCC) in Florida, the shuttle Mission Control Center (MCC) in Houston, and the Space Telescope Operations Control Center (STOCC) in Maryland. We five would operate out of the small "crew quarters" living space on the Kennedy Space Center campus, where astronauts since Gemini days had spent their final nights before leaving Earth. The following two days would replicate all the events of the day before launch (abbreviated as L-1, "L minus one," in NASA parlance) and launch day. There were some commonsense deviations from launch day reality, of course. Our timeline would be run during normal business hours, rather than starting before dawn, as it would for the real launch. More importantly, the shuttle was not fully fueled, no rocket engines would fire and the shuttle would remain bolted firmly to the pad this time.

Loren and Charlie flew their final round of landing practice early Monday morning, just as they would on the morning before liftoff. The normal L-1 day briefings on the weather outlook for launch and status of shuttle systems came next, followed by a final inspection of our launch and entry suits. We filled our afternoon with two other important events: a round of visits with the ground crew who had worked over the previous six months to get our spacecraft ready for us, and the long-awaited trip to the launch pad to inspect Hubble in *Discovery*'s payload bay.

As was customary for these prelaunch visits, we were given several hundred mission emblem pins to hand out to members of the Kennedy Space Center ground crew as thanks for all the hard

FIGURE 7.6

STS-31 “Lunch Team” pin. Source: Author’s personal collection.

work they had done to get *Discovery* ready to fly. At first glance, the pins looked very sharp. In a nice touch, a small crescent with the words "Launch Team" had been added to the bottom of our oval patch. Unfortunately, closer inspection revealed that we had actually been given several hundred "Lunch Team" pins (figure 7.6). This, of course, only made them a bigger hit with *Discovery*'s ground crew.

✷

It always sent chills down my spine to finally see the payload I had worked on for months, if not years, bolted into the cargo bay of a space shuttle, on the pad, and poised for launch. Everything that had sometimes seemed distant and abstract amid all the meetings, paperwork, mockups, prototypes, and simulations was at last truly, vividly real. Nestled snugly in *Discovery*'s payload bay, Hubble again reminded me of a beautiful silver gift from Tiffany's, just as it had when I first saw it in Sunnyvale five years earlier. We climbed along the launch pad support structure that enveloped Hubble and the payload bay to get a closer look at our precious cargo. This would be our only opportunity to study a number of items that were critical to our mission but would be difficult to see from the shuttle's flight deck on deployment day.

Bruce and I were particularly focused on the umbilical cables connecting the telescope to the shuttle's power supply. If the automatic disconnect mechanism failed to work properly, we would end up doing a contingency spacewalk to release them by hand. The tight fit of the telescope in the payload bay was another area of concern, especially to Steve. There were just inches to spare between the wider aft section of the telescope and the side of the payload bay. We all understood intellectually that it was a tight fit—both engineering data and simulations had made that clear—but seeing it up close was something else. Charlie was stunned to find that the gap was barely wider than his fist. That really drove home how small a margin for error Steve would have in the early stages of unberthing the telescope with the shuttle's robotic arm. We left the pad hoping that we would next see Hubble through the shuttle's aft flight deck windows, glinting in the sun and framed by the beautiful Earth below.

✷

Launch day (simulated). Wake up, shower, hop into a flight suit. Check in with the flight surgeon for my final medical clearance. Head to the dining room for a bite of breakfast (remember, this will be the traditional "happy crew, raring to go" photo op on the real launch day). Cross the hall into the conference room for the final weather update and status briefing with the launch team. Everything's GO; excellent. Time to suit up. Back to my bedroom. Take a final potty break, put on the fleece shirt and leggings I wear under my pressure suit. Pack my earthling clothes and everything I'm not taking aboard the shuttle into the large duffel bag, tagged with the mission number and my name, and leave it on the bed for the support crew. On to the suit room. Wriggle my legs and arms into my launch and entry suit through the back opening. Push my head through the neck seal, then stand up to pop my hips inside. Pull the zipper from the back of my neck down through my crotch and lock it in place. Lace up my boots. Check that my flashlight, pencils, and pocket knife are all in the correct pockets. Don the brown-and-white "Snoopy cap" that keeps the radio microphones near my mouth (so-called because it looks like the headgear worn by Charles Schultz's Snoopy as the World War I Flying Ace), then attach my gloves. Slip my helmet over my head and lock it onto the neck ring. Now sit back in the suit room's comfy recliner and relax, while my suit techs go through their checkout procedures and we all wait for word that we're GO to depart for the pad.

Though this time we will be back in just a few hours, the hallways from the crew quarters to the waiting astronaut transfer van (a modified Airstream motorhome) are lined with people seeing us off: the secretaries and suit techs, medics, and support team who have made the long journey to this spaceflight with us, along with Kennedy Space Center employees whose offices just happen to lie on our path. It's a lighthearted, happy walk on this rehearsal day (figure 7.7). The mood will be more serious the next time we walk this route, when just a few hours remain in the countdown to a real launch.

The five of us each take one of the recliners aboard the astrovan and connect our suits to a small air conditioning pack to stay cool on the ride out to the pad. We are joined by chief astronaut Dan Brandenstein and Don Puddy, who replaced George Abbey as the head of flight crew operations. This is a full dress rehearsal for them, too. Just like on launch day, they will ride with us as far as the launch control complex, then hop off and head to their designated stations. Lighthearted banter sprinkled with jokes helps pass the time as we drove. We drop Dan off first, then stop at the gate to the launch control center to drop off Don. As the door closes behind him, the five of us joke about our bosses bailing out on us as we draw closer and closer to the live explosives on the launch pad.

Only the five of us left in the van now, plus the driver and a couple of suit techs. Nobody goes any closer to the pad than the launch control center when a fully fueled space shuttle sits on the pad, except for the folks who ride rockets for a living—that would be us—and the small squad of technicians who will strap us into the cockpit and seal the hatch. Then those sensible folks will fall back to the launch control complex, leaving us alone at the very center

FIGURE 7.7

Leaving the crew quarters for our countdown rehearsal. Front row: Charlie Bolden (left) and Loren Shriver (right). Middle row: Bruce McCandless (left) and Kathryn Sullivan (right). Third row: Chief Astronaut Dan Brandenstein (blue flight suit), Director of Flight Crew Operations Don Puddy (dark suit), Steve Hawley (right). Source: NASA.

of the three-mile diameter "blast danger" zone. We will be very happy to be there.

The van makes its final turn, passes through the pad perimeter gate, and starts climbing the sloping ramp that leads to the surface of the launch pad. On our real launch day, we will crowd forward to the windshield and strain to catch a glimpse of our spaceship, gleaming brightly against the predawn darkness under the pad's bright xenon flood lights. We won't see *Discovery* on this rehearsal day, since she is still cocooned inside the pad support structure. "I hope this is the last time we drive off the launch pad," someone says (probably Charlie or Steve, who had endured five scrubs before finally launching on STS-61C in early 1986). "Next time we come out here, we're leaving straight up."

8

A NEW STAR IN THE SKY

There we were, all dressed up and strapped in tight, but going nowhere. On this day, April 10, 1990, we would climb out of the shuttle and head back to our quarters instead of rocketing into orbit. The five of us onboard the shuttle and everyone in the control centers at Kennedy and Johnson knew this fully four minutes before our scheduled liftoff time, when one of the shuttle's auxiliary power units failed to start properly. Despite this glitch, the countdown clock kept ticking down, leaving the crowds gathered along the Florida shoreline to believe for a little while longer that they would see a shuttle launch today. They wouldn't learn the disappointing truth until the clock stopped with thirty-one seconds to go, the last preplanned stopping point in the countdown timeline. Then they would head dejectedly to the beach or to their hotel, while the launch control team went through the procedures needed to back the space shuttle carefully away from the brink of liftoff. The small silver lining for the five of us was that the turnaround to another launch attempt was sure to take at least ten days, meaning we would be let out of quarantine and able to visit with the friends and family who had come to the Cape for the occasion.

We flew back to Houston the next day, still awaiting final word on how long a delay we were facing. Fourteen days was the answer: The next launch attempt would be on April 24. Getting back to peak readiness after a launch scrub was a familiar challenge for Charlie and Steve. Steve's first launch attempt in 1984 had ended with the only on-pad abort in space shuttle history. The shuttle's three main engines started on schedule, six seconds before liftoff time, but were quickly shut down when the onboard computers detected a problem. His quip, "Gee, I thought we'd be higher than this at MECO [Main Engine Cutoff]," was widely hailed as the best one-liner ever delivered by a TFNG. His next mission, with Charlie as second in command, scrubbed six times before finally getting off the ground in January 1986. A lot of slightly nervous joking about "the Hawley launch curse" filled the idle moments of our second stay in quarantine.

Coming back from a launch scrub may have been old hat to Charlie and Steve, but it was new territory for me, and I found the ground quite uncomfortable. When we went out to the launch pad the first time, I knew I was one hundred percent ready to fly. Now I had to figure out how to make sure that was true again in fourteen days' time. Was it best to just chill out, or should I cram in more sim time or briefings? I chose the chill path. Happily, the short slip gave us very few working days to fret about this. We went back into quarantine one week to the day after the launch scrub and flew back to Florida four days later.

The Hawley curse seemed to still be with us when the countdown clock stopped at T-31 seconds again on April 24. The culprit this time was a valve on one of the fuel lines that fills the external tank. According to the data on the propulsion engineer's console, one of the two valves installed to close this line was still open. If that was true, then any fault in the other valve would let fuel leak overboard instead of feeding into the shuttle's three main engines. The potential consequences of a main propellant leak were all enormous. Ending up in an orbit too low to deploy Hubble was the least bad scenario. The more calamitous prospects ranged from aborting to an emergency landing runway on the coast of Africa to bailing out somewhere over the Atlantic Ocean. We would scrub rather than accept that risk. On the other hand, perhaps the indicator on the console was wrong. Maybe the valve itself was fine but the instrumentation that fed data to the engineer's console was bad (even spaceships can suffer from flaky sensors). That would mean the engine system was fine and there was no reason to scrub. Which was it: Serious problem, or faulty indicator? Go for launch, or scrub?

This high-stakes call fell to the main engine expert in the launch control center, known by the call sign MPS (Main Propulsion System). Time was not on his side. Only twelve minutes remained in the prelaunch fuel budget for our auxiliary power units. We listened intently as the launch team analyzed the problem. "MPS, what's your status?" the launch director asked. The propulsion engineer talked calmly through the data on his display and the principles of basic physics that convinced him the valve was indeed closed. He re-sent the command to close it, hoping this would correct his console reading. That worked, but the control center computers still had a lock on the countdown clock. "MPS, what is your call?" the launch director pressed. "I am prepared to manually override the ground software and proceed with the count," he replied. The launch director gave him the GO to do that and alerted the other launch controllers to get ready to continue the countdown. Minutes later we heard the call we all (but especially Steve) had been hoping for: "All controllers, this is NTD [NASA Technical Director]. The countdown clock will resume on my mark. Three, two, one, MARK." The entire episode had taken just two minutes and fifty-two seconds.

Thirty-one seconds later, *Discovery* blasted off the launch pad. I closed my eyes and took in the sounds and sensations of a space shuttle launch. For the first two minutes and fifteen seconds, the ride was turbulent and loud, like being in a wild combination of earthquake, rock concert, and fighter jet. The vibrations were almost bone-rattling; the thrust pushing upward through my back strong and constant. I felt the thrust tailing off, heard Charlie report that the solid rockets were burning out as expected, and then heard the *thump* that announced they had been jettisoned. Now the ride seemed quiet, and as smooth as an electric train. The push against my back continued, as the main engines accelerated us toward orbital velocity. When the engines cut off six minutes later, the lightness in my limbs and the checklists floating at the end of their tethers confirmed that I was back in orbit. I felt instantly at home.

The first order of business after main engine cutoff was to shift the shuttle's systems from launch mode to orbit mode and unpack the equipment needed for routine operations. With that complete, we turned to the main objective of our first day in orbit: getting ourselves, our EVA equipment, the shuttle, and Hubble itself ready for the following day's deployment activities. Bruce and I donned our oxygen masks to start washing the nitrogen out of our bodies and set about unstowing our tools and checking out our spacesuits. Above us on the flight deck, Steve and Charlie powered up and checked out the shuttle's robotic arm, then made a thorough visual inspection of the telescope in the cargo bay.

We ended our first day in orbit in high spirits. The robotic arm was working well, the telescope showed no sign of damage from the harsh forces of launch, and our spacesuits checked out perfectly. The last event before we bedded down for the night was a routine check-in call with the flight surgeon in mission control. This one drew heightened attention because of the next day's high-stakes operation and the potential for a contingency spacewalk. Such momentous events were not normally scheduled for the second day in orbit, in case a crewmember was still adapting to microgravity.[1] Since we were all feeling fine, Loren turned the conversation to another topic: we wanted the engineer who had saved the launch to be at Edwards Air Force Base when we landed, so he could join in celebrating a successful mission and we could thank him publicly for his pivotal role in making it happen. There was nothing our poor flight surgeon could do to make this happen, other than relay our demand as forcefully as possible to the higher-ups at mission control. Loren extracted the surgeon's pledge to do just that before signing off for the night.

✷

Hubble deploy day for us began with an entirely forgettable wakeup tune and the routine start-of-day tasks: dress, grab a bite of breakfast, review the morning message package, update the shuttle's navigation data, and make sure all of our middeck experiments were running properly. The ground segments of Team Hubble were gearing up for their big day as we went about our morning tasks. At Goddard, the Space Telescope Operations Control Center was sending commands to check out the telescope's onboard systems. The shuttle mission control team, along with the engineering experts from Marshall and Lockheed who had come to Houston to be on hand for the mission, were following the checkout closely and ensuring that all the shuttle's systems were ready as well.

With our morning's shuttle housekeeping chores done, we fanned out to prepare the equipment we would need for deploy operations. Bruce and I put on the long-johns-style cooling garment we would wear under our spacesuits and staged all our equipment in the airlock and middeck so that no time would be lost if an EVA became necessary. Then we joined Loren, Charlie, and Steve on the flight deck for the main event. Our roles were by necessity limited to photo documentation and lending the occasional helping hand, since any hiccup in the deployment could pull us away to suit up and head out on an EVA. I had claimed the job of chief photographer, and now had every camera and lens we owned arrayed within easy reach on the aft flight deck. We were midway through our fifteenth lap around the planet, aiming for deploy on the nineteenth lap, in just under six hours.

Steve grappled Hubble with the shuttle's robotic arm and Charlie released the large, sturdy latches that secured it into the payload bay (figure 8.1). With mission control's GO, we switched the telescope over to internal battery power and disconnected the shuttle's electrical umbilical. This started a crucial clock: both solar arrays had to deploy successfully within just over six hours, or the telescope's batteries would run down and die. A huge amount of pre-mission planning and training had been devoted to ensuring there was no fat in the deployment timeline and preparing the combined Johnson-Marshall-Goddard-Lockheed team to operate efficiently. Now it was showtime.

"Plans are nothing, planning is everything" is a common adage among astronauts. We cite it frequently to remind ourselves that unforeseeable events and circumstances virtually guarantee that even the most perfect plan will not work as intended in the real world. The carefully crafted deployment plan began to fray as soon as Steve started to lift the telescope out of the cargo bay. Instead of rising steadily upward on the end of the arm, as it always had in the simulators, it began to pivot left and right. Hubble filled the cargo bay like a loaf of bread fills a baking tin—with no

FIGURE 8.1

View from *Discovery*'s aft-facing windows as Steve Hawley grapples Hubble with the shuttle's robotic arm. Source: NASA.

room to spare, and certainly no room for unexpected movements like the ones we now saw out the windows. Terrified at the thought of bumping this delicate instrument against the side of the orbiter, Steve quickly put on the brakes and called mission control. They agreed with his assessment that the software controlling the robotic arm's automatic mode was likely causing the problem and that it would be best to switch to manual mode, even though this would cost precious extra time. Using a toggle switch to command each of the arm's joints one at a time, Steve began to inch the telescope out of the cargo bay. Once it had cleared the sides of the bay completely, he went back to auto mode and maneuvered the telescope to the position required for deploying the solar arrays and antennas. We were thirty minutes behind the timeline when he got there.

The game plan for deploying the telescope's appendages, as the solar arrays and antennas were called, began to unravel next. Almost every step in this plan revealed that the ground control team at the Space Telescope Operations Control Center had not fully appreciated the complexity of Hubble's systems and was struggling to deal with the pace and stress of real-world spaceflight operations. With the exception of the three joint integrated simulations back in March, their training had not included the kinds of challenging scenarios that force a team to confront unexpected events and develop the troubleshooting and coordination skills needed to solve them efficiently. They lost valuable time struggling to reach decisions on what should have been minor issues, like why an appendage moved slightly when its launch latch was released. Secondary symptoms of their difficulties were clear to us aboard *Discovery*: repeated changes to the planned sequence of activities, slow decision making, the occasional hint of exasperation in Capcom Story Musgrave's voice and, most worrying, the rapidly shrinking amount of time left on Hubble's batteries.

After lots of troubleshooting, more juggling of events, and yet another long wait, this time for the right lighting conditions, we finally got the GO to unfurl the solar array wings. The five of us crowded around the aft flight deck windows, cameras at the ready and all eyes trained on the telescope. The Hubble control team was counting on us to watch this delicate operation very closely, since we would see any hiccup in the process precious moments before engineering data reached their consoles. Mission control reported that the command to unfurl had been sent to the first array. The spreader bar moved away from the boom smoothly, pushed by the thin bright silver tubes along each edge of the blanket. The twin wings sprouting smoothly from the solar array mast reminded me of the first plants of spring emerging thorough the soil. It was a stunning sight, and the glowing golden wings made Hubble more beautiful than ever.

The motion stopped at exactly the predicted time. One solar array deployed, one to go. The space telescope operations team regrouped quickly and sent the deploy command to the second array. Again, the spreader bar moved right away, but this time it stopped with just a foot or so of the array unfurled. Bruce and I thought for sure that mission control would direct us to start preparing right away to go outside and drive it out manually. With the tight limit on battery time in mind, we were spring-loaded to head for our spacesuits at the very first sign of trouble. Instead, there was silence, as the telescope control team deliberated. A second deploy command was sent. Once again, the array began to move. Once again, it soon stopped. About 20 percent of the blanket was now unfurled. Bruce and I weren't waiting for instructions this time. We dove down to the middeck to start suiting up.

Charlie joined us a few minutes later to help with our suit checks and go through the steps needed to depressurize the airlock. Bruce and I slipped into the pants-like lower pieces of our space suits, and then took turns floating into the airlock and wriggling ourselves into the upper torso sections mounted on the airlock wall. Charlie hooked up the hoses that fed chilled water into the plastic tubes laced into our long underwear, then closed and locked the seal between the two segments of each suit. Our Snoopy caps came next, followed by our gloves and then helmets (figure 8.2). Every time he had buttoned us into our suits during training, we had been surrounded by an ocean of air and dozens of technicians and medics ready to respond to any emergency. This time it was just the three of us, and two of us would soon head out into the vacuum of space. Charlie's always

FIGURE 8.2

The author during a short break during suit-up aboard *Discovery*. Source: NASA.

high level of thoroughness and care went up a couple of dozen notches as this reality hit him.

The amount of nitrogen still in our bloodstream made it unsafe for us to just rush outside, as much as we wanted to do that. Buttoned up tightly into the pure oxygen environment of our spacesuits, we settled in reluctantly for yet another round of prebreathe. Forty minutes later, Charlie came back to do a final inspection of our suits, unhook our backpacks from the airlock wall and close the hatch separating us from the shuttle cabin. His next step, after getting a "GO for airlock depress" from mission control, was to open the valve that allowed the air inside the airlock to escape into space. The protective cap covering the valve proved very hard to loosen, holding us up for about thirty minutes. Finally Charlie opened the valve and dropped the airlock pressure to five pounds per square inch, halfway to vacuum. We stopped there for our final safety checks. Both suits checked out fine. Bruce and I were ready to go. All we needed now was a "GO for EVA" from mission control.

What we got instead was "Stand by." Analysis of telemetry from the stuck solar array showed that it had stopped because of a sensor reading that said the tension on the blanket was too high. All the while we had been suiting up and prebreathing, the telescope control team had been working to determine whether or not the tension reading was valid, and what to do with the array if it was. Mission control wanted to give them a little more time to work the problem before committing us to an EVA.

Jammed together inside the airlock, Bruce and I waited in silence. There was nothing to look at but the stark white airlock wall that filled my visor. The sounds of my suit filled the pauses between radio transmissions from the ground and Loren and Steve's conversation on the flight deck. Because of the way our safety tethers were routed, it would be my job to crank out the jammed array if we ended up outside. I spent our idle time visualizing every move I would make once the outer hatch opened: slide out of the airlock, hook up my safety tether, move down to the toolbox at the bottom of the cargo bay, attach the tool caddies and portable foot restraint I would need, work my way up to the port sill of the cargo bay, and head aft until the telescope was right above me. When Steve lowered it to within my spacesuited arm's reach, I would grab the lowest handrail, make my way up to the base of the solar array, and go to work.

I was excited at the prospect of doing another spacewalk, but anxious at the same time. I was duly impressed by the thought that Hubble's fate might end up in my hands. That was, after all, precisely why we were here. Bruce and I were the only two Hubble repair spacewalkers in the universe. We had spent hundreds of hours testing our tools and learning every detail of the telescope that was now parked above us on the end of the shuttle's robotic arm. We knew our tools and procedures cold, because we had built them and written them, working hand in glove with the engineering and EVA teams from NASA and Lockheed. We had done everything we could to be ready for this moment, hoping all the while that it would never come. But here we were, about to put all that work to the ultimate test 330 miles (611 km) above the Earth, with NASA's flagship scientific mission hanging in the balance. The plea commonly called "the astronaut's prayer" flashed through my mind: *Please don't let me screw up.*

After what seemed like a very long time, mission control informed us of the new plan: the team at the Space Telescope Operations Control Center had concluded that the blanket tension sensor was faulty and was sending a command to disable this check in the onboard software. As soon as that was done, they would reissue the command to unfurl the array. The goal was to release the telescope into free flight around orbital noon on the upcoming daytime pass. Not only would there be no EVA, but Bruce and I would not even see the deployment, since we were to remain in the airlock until the telescope had been released and the shuttle had separated from it to a safe distance. This was not what I had hoped to hear, but it made perfect sense. The ground team had wisely chosen the fastest and least-risk path to completing the telescope's deployment. Quite rightly, our desire to watch didn't enter the equation. On the shuttle's intercom loop, Charlie apologized for leaving us stuck in the airlock and headed topside to assist Loren and Steve with the deployment.

I followed the rest of the events through the flight deck chatter on the intercom and air-to-ground communications from mission control. The stuck solar array unfurled without a hitch when the telescope control team commanded it again. Steve then maneuvered the telescope to the release position, and Loren commanded the shuttle into its proper attitude. A short silence followed, which surely meant mission control was polling all the flight controllers to ensure that everyone was ready to set Hubble free. The Capcom soon made the call we had been waiting for all day: "*Discovery*, Houston, you are GO for HST release." Steve opened the grapple fixture and backed the robotic arm slowly away from the telescope (figure 8.3). Hubble and *Discovery* were now two independent satellites, flying just a foot or so apart. A moment later, I felt the bump that announced that Loren had fired the engines

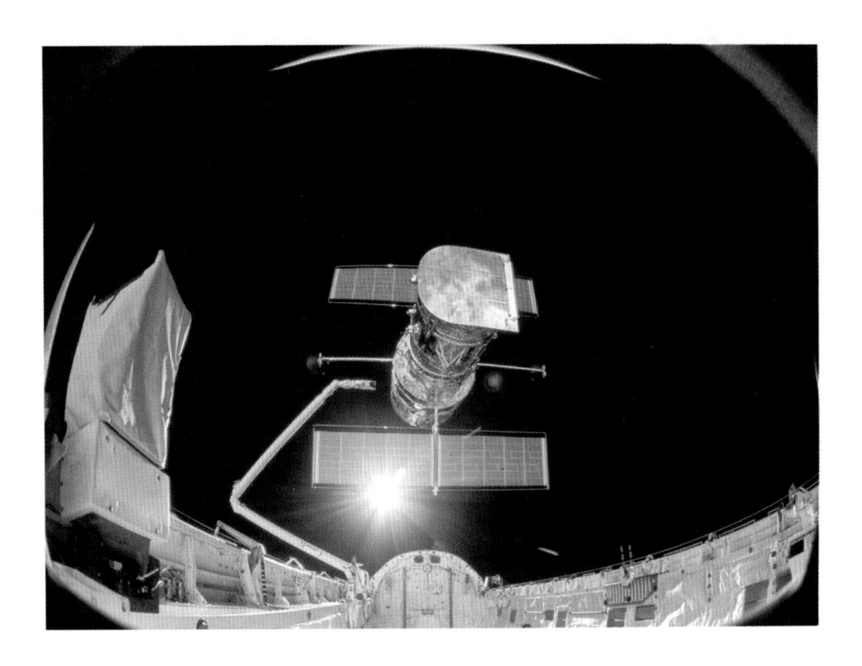

FIGURE 8.3

Hubble and *Discovery* mere inches apart, moments after Steve Hawley released the shuttle's robotic arm. Source: NASA.

to separate us from the telescope. It had been easy to envision all the technical actions that had led to this moment, but I could not imagine how beautiful a sight it must be to behold. I envied my crewmates on the flight deck, who were chattering boisterously as they took photo after photo of Hubble against the backdrop of Earth and space. Eventually, Charlie pulled himself away from the flight deck windows and came back down to the middeck to get us out of the airlock. The telescope I had likened years before to a creation by Tiffany's was now the newest artificial star in the sky.

✷

We spent the remaining three days of the mission in a station-keeping position about fifty miles away from Hubble, ready to return to it for repairs or retrieval if a major problem turned up during the initial checkout of the telescope's operating systems. Life aboard *Discovery* was almost leisurely for us, since the telescope's size and weight had left us very little room for other scientific equipment. The few experiments we had inside the cabin and general shuttle housekeeping chores set our daily pattern but left plenty of time to watch the Earth slide by below us. Thanks to Hubble, we were flying roughly twice as high as the normal shuttle altitude, so the view was more stunning than ever. One moment that stands out in my memory is a night pass across the Amazon basin and the Caribbean that I watched with Charlie. Lightning flashed and rippled continually within the large thunderstorm complexes below us. To people on the ground beneath those storms, the lightning would seem episodic, coming and going as individual storm cells passed overhead. From our vantage point, the activity was continuous, with a bolt or horizontal streak of lighting flashing somewhere in the storm complex all the time. Charlie likened the magical sight to watching a concert light show with the sound on mute. I've always wished someone would compose the music we couldn't hear that night.

Discovery touched down at Edwards Air Force Base just before 7:00 a.m. Pacific Standard Time on April 29, 1990. We had traveled over two million miles on our eighty laps around the planet, placed a new star in the sky—as bright as the star Spica—and come home safely.[2] When the moment came to address the landing team and others who had gathered to welcome us to Edwards, we bounded onto the platform, grinning like fools and waving happily to the crowd. Our flight surgeon had delivered on his pledge: the propulsion system engineer from the launch control center was front row center in the crowd, grinning every bit as widely as we were. I loved these postlanding moments. I always felt as I imagined Olympic athletes must feel, when years of training and sacrifice are swept up in the emotional whirlwind and the brilliant glare of a long-imagined gold-medal celebration. Just like Olympians, we struggled to find the words to share even a little bit of our experience with our support team and to not choke up as we thanked them for all their hard work and dedication. I failed on both counts every time.

9

RESCUE AND RENOVATION

Our high hopes for a spectacular first image from Hubble came crashing to Earth a few weeks later, when the world learned that the multibillion-dollar space telescope we had just put into orbit had blurry vision. Charlie and Steve spent many long weeks worrying that they might have caused this by bumping the telescope as they lifted it gingerly out of the shuttle's cargo bay. They must have been the only two people on Earth who were relieved to learn that Hubble's 94-inch-diameter primary mirror was the wrong shape. It was too flat at the perimeter by 0.0001 inch (0.003 mm), which is about 1/50th the diameter of a human hair or 1/40th the thickness of a typical hardcover book page.[1]

This was unbelievable news, an unthinkable error. A tidal wave of shock and anguish swept over NASA and the Hubble science community. Congress and the media erupted in outrage. "It was as if an eagle had become a bat," wrote Arthur Fisher in the October issue of *Popular Science*. The pain was written clearly on the ashen faces of the NASA officials who broke the news to the public. The crippled telescope quickly became the newest metaphor for incompetence and technological hubris, ridiculed in print, by virtually every late-night talk show host, and on the silver screen (figure 9.1). Some pundits linked the Hubble foul-up to the mistakes that caused the loss of *Challenger* and cast it as the death knell for a NASA that had long since lost its way. Congress followed hot on the heels of the comedians and pundits, convening public hearings at which they grilled senior NASA leaders mercilessly.[2]

I watched in dismay from the sidelines as all this unfolded. There had been no going to the back of the waiting line for me when we landed, since I had been given another flight assignment several months before launch: Payload Commander for a Spacelab mission devoted to upper atmospheric research, known as ATLAS-1. As Payload Commander, I would lead the crew's four-member scientific team, consisting of myself, fellow mission specialist Mike Foale, and two "payload specialists," as NASA called any non-NASA person on a crew. In our case, these would be two highly qualified scientists, Byron Lichtenberg and Michael Lampton, whose deep expertise with our scientific instruments would be invaluable to the mission.[3] Training for my new payload crew had begun even before STS-31 launched, so I had to hustle through my postlanding duties to join my new crewmates as soon as possible.

Bruce was my primary source of Hubble news through the spring and summer of 1990. He was in the thick of things as usual, having pulled some very high-level strings to get himself named to the team of engineers and scientists who were working feverishly to figure out what could be done to fix Hubble's blurry vision and salvage as much science as possible.[4] On the face of it, this seemed an assignment straight out of *Mission Impossible*: "The eight-foot diameter mirror that is orbiting 330 miles (611 km) above the Earth

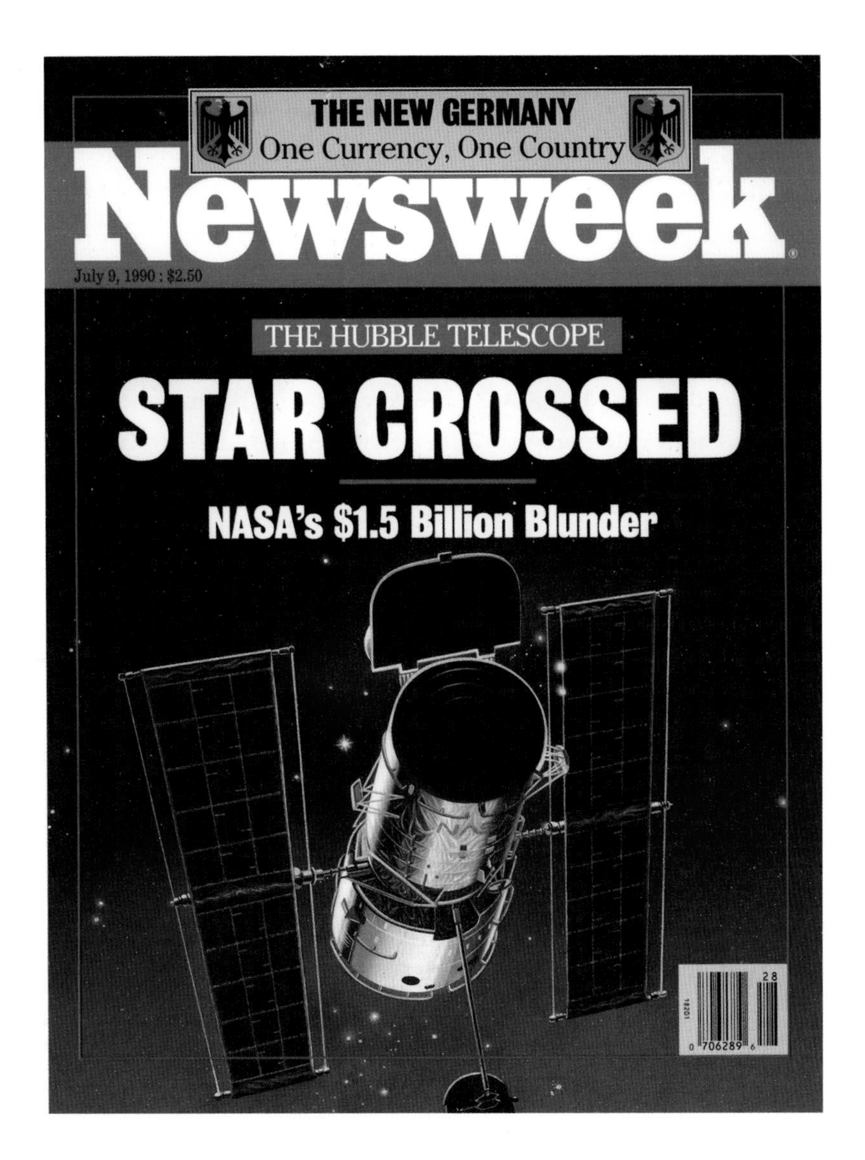

FIGURE 9.1

Newsweek magazine cover, July 9, 1990. Source: Artwork courtesy Daniel Kirk.

is a teeny-tiny bit misshapen. Your mission, should you choose to accept it, is to fix the mirror and save the scientific mission."

As is often the case, the trick to finding a solution was to recast the problem.[5] Did the mirror itself really need to be fixed? Suppose instead that the real challenge was to correct the light that it reflected into the instruments? One happy circumstance—the lone little sliver of good news in an otherwise truly horrible fiasco—suggested that this might be possible. The mirror's shape was indeed wrong, but it was precisely wrong. This meant the engineers could calculate very precisely the difference between its actual shape and the shape it was meant to be. This information could then be used to calculate the correction that would restore the telescope's vision, much as an optometrist determines the shape of the prescription lenses needed in a pair of eyeglasses. By October, the team had the outlines of a recovery plan in hand. One of the four large scientific instruments in Hubble's aft end would be replaced with an identical box containing the corrective optics—small mirrors in Hubble's case, rather than lenses. These mirrors would provide properly focused light to the telescope's guidance sensors and the three remaining instruments. Similar mirrors would be built into the replacement unit for Hubble's main imaging instrument, the Wide Field/Planetary Camera, which was already under construction.

Launching the first maintenance mission two to three years after deployment had long been the plan, but nobody had ever imagined that the Hubble mission's life would be hanging in the balance the very first time out.

✷

It felt very odd to be outside the Hubble world. After five intense years of work to build a solid foundation for the critical maintenance mission that lay ahead, there was suddenly nothing I could do to ensure its success. Parents must feel like this when their children leave home, but at least they get emails, phone calls, holiday visits. My split with Hubble was cold turkey. I ached to know how the new mission planning team would build on the foundation we had bequeathed them. Where would it prove to be as solid as we intended? Where would they discover gaps? When would they realize that the maintenance tasks required more spacewalking time than the shuttle claimed it could provide, and how would they solve this show-stopper of a problem? How would Goddard actually handle this huge new responsibility? How much knowledge would be lost as our deployment mission team moved on to other assignments and the Goddard servicing team took over? I could only wonder.

I could also pester Ron Sheffield, and did so frequently. He and his two key lieutenants—Brian Woodworth and Peter Leung—had stayed with Hubble and were already vital members of the newly formed servicing mission team. The three of them had inspected every inch of the telescope countless times over the preceding five years. They had fit-checked every EVA tool on every bolt and connector, closely observed the removal or reinstallation of dozens of electrical and mechanical components, captured fine-scale details of every operation from the telescope system engineers, taken tens of thousands of pictures of the inside and outside of the telescope, documented every technical specification and test result, and compiled all of the above into a database. Nobody else came close to matching their understanding of Hubble as a maintainable system or their skill in navigating the multiple NASA centers involved in Hubble missions. They would remain the mainstays of Hubble servicing for nineteen years, teaching and supporting every Hubble servicing mission crew.

CHAPTER 9

✷

War in the Persian Gulf was looming as I took on my first command as a Navy officer in October 1990. My unit was a small, specialized outfit based at the Naval Air Station in Dallas that provided meteorologists and oceanographers to the Oceanography Command Center in Guam. Every US Navy ship and airplane operating in the Persian Gulf region relied on this center's weather, shipping, and combat operations forecasts. Operation Desert Shield, the United States' initial military response to the Iraqi invasion of Kuwait, had already pushed the center to its maximum operating capacity. More sailors would clearly be needed if open warfare broke out. My first order of business as the new commanding officer would be to prepare my unit to be recalled to active duty.

The oceanography center's skipper, Captain Dieter Rudolph, convened his first war cabinet meeting two hours after my plane touched down in Guam on Sunday, January 13, 1991. I introduced myself to him and took a seat in the small top-secret conference room. Events in the Gulf were moving much more rapidly than had been apparent back home. The captain expected that the order to execute the war plan would come down within days. He signaled me to walk with him as we left the briefing. "I need to keep you here as we activate your unit," he said. "I'm extending your orders." With that, my plan for seventeen days of duty in Guam blew up.

Operation Desert Storm began four days later.

Despite my active duty military status in Guam, the war was largely a television experience for me, just as it was for most Americans. The center had split its operations team into two squads, one working in a top-secret area and the other in the usual workspaces. The classified group provided combat forecasts for the naval operating forces in the war zone, while the unclassified group supported general operations in the western Pacific and Indian Oceans. I spent my days supporting the latter group and working on the recall of my unit via phone and email with my XO back in Dallas (the executive officer and second-in-command of our unit). My sailors would augment the unclassified group when they arrived in Guam, allowing the center to move more of its full-time sailors onto the classified squad.

The recall was in full swing when I briefed Captain Rudolph in early February. Our top-priority people were sorting out their personal affairs in preparation for an indefinite deployment to Guam. They would be on-island within a few weeks. Paperwork was underway to recall a second wave of sailors, to ensure that the center could sustain the heightened operating tempo throughout the war, however long that might turn out to be. That left the question of my status. The captain had extended my original seventeen-day orders to thirty days and now wanted me to stay for the duration of combat operations. Luckily for me, the president had only authorized a *voluntary* recall, so the captain could ask but not order me to stay. I had been away from my ATLAS-1 crew for almost a month, and we were about to shift into full-time flight training. It was time for me to go home.

*

Fast-forward to early January 1992, coming up on launch minus ninety days for STS-45 and the ATLAS-1 atmospheric science Spacelab mission.

We were now second in line for the launch pad, behind another Spacelab flight devoted to materials science experiments. Next in line after us was STS-49, a satellite rescue mission. Everyone around the space center was looking forward to that exciting mission and the dramatic spacewalks it would feature. No such excitement surrounded our mission. Scientists punching computer keys are simply no match for the spectacle of spacewalking astronauts snatching satellites out of the sky. The joke around the astronaut office was that our mission was merely a speedbump on the road to the excitement of STS-49. There sure were days when it felt like that.

I picked up snippets about preparations for the Hubble servicing mission from the brief reports that Story Musgrave, then the EVA lead for the office, provided at each Monday morning's All Astronauts meeting. A lot had been accomplished since I had gotten swallowed up by STS-45. The launch date was holding firm for late 1993. The list of maintenance tasks to perform in addition to the optical fix for the flawed mirror had been nailed down. As our M&R team had predicted back in late 1989, it would take far more than two spacewalks to get it all done. Driven by both the urgency of fixing Hubble and the looming challenge of building a space station, the shuttle team had managed to increase the number of planned spacewalks per shuttle mission from two to five. The rough choreography for maintenance mission spacewalks we had developed prior to STS-31 had also been refined substantially, through several rounds of neutral buoyancy testing involving over a dozen astronauts. The buzz around the office was that at least one veteran spacewalker would be assigned to the mission shortly, given the complexity of the spacewalks involved and the amount of work still ahead.

With my own flight training about to shift into top gear, I scarcely had time to think about the Hubble mission or any other future flight. In fact, what little time I had to think about anything other than STS-45 was filled with other subjects altogether. Our home planet was the first of these. My Navy duty in Guam, the environmental damage done by the Gulf War, and our scientific flight training had all brought my lifelong fascination with Earth back to the foreground of my thoughts. I wasn't sure what to make of this reawakened interest. Perhaps it signaled that the astronaut chapter of my career was ending. That thought raised a raft of big questions that I couldn't begin to tackle during the final sprint to launch. I put all such thoughts and questions into a little mental jar labeled "Do Not Open Until After Landing" and tucked it away for safekeeping.

A phone call from Her Deepness, marine biologist Dr. Sylvia Earle, broke open my mental jar before the month ended.

*

Sylvia and I had first crossed paths in 1981, when we became the first women to join the renowned Explorers Club. She gained the moniker "Her Deepness" after a 1979 solo dive to 1,250 feet (381 m) in the waters surrounding Hawaii, during which she walked along the seafloor wearing a kind of diving armor known as a JIM suit. At the time of her telephone call to me, Sylvia was serving as the chief scientist of the National Oceanic and Atmospheric Administration (NOAA), a post I had encouraged her to take when the first Bush administration approached her in 1989.[6] She called to tell me that she was stepping down to take care of some family matters and wanted to recommend me as her successor. Would I agree to throw my hat in the ring?

This prospect was tantalizing. NOAA's mission, combining scientific research with practical applications, was a great match for the mix of curiosity and pragmatism that had always driven me. The agency's research and operations included all of my favorite topics and toys (oceanography, meteorology, and cartography; satellites, ships, and aircraft), promising me both a strong starting foundation and ample room for further learning. Its ties to industry, academia, and other parts of government would make the position a great perch from which to explore those sectors as potential future career destinations.

Leadership opportunity topped the list of the job's attractions. My Navy command assignments had allowed me to start honing my leadership skills, and I was eager to continue this. The chances of doing so in the flat organizational structure and cliquish culture of the astronaut office seemed extremely slim, and the broader NASA world did not look much better. I could count the number of women then holding senior leadership positions on one hand and still have several fingers free. Topping out as some man's deputy or senior staff aide seemed to be the rule. The NOAA prospect would catapult me to the number three position in a four-billion-dollar federal agency, and no doubt open the door to further leadership roles. Provided I succeeded, of course.

With the blessing of my ATLAS-1 mission commander, Charlie Bolden, I flew to Washington, DC, for an initial interview with NOAA Administrator Dr. John Knauss. John, a good salesman with a deep passion for NOAA's mission, made the job sound even more interesting than I had imagined. I left the meeting knowing I was on his short list of names to send forward to the White House, and that he would get back to me when he had a final decision. A lot of still-unanswered questions swirled through my head on the flight home. Once again, I tried to stuff them all into a mental jar and turn my full attention to our final weeks of training.

*

The office buzz about an early Hubble servicing mission assignment became reality in early April with the announcement that Story Musgrave would be the flight's Payload Commander and lead spacewalker. A veteran of four space shuttle missions at the time, Story had played a major role in getting the shuttle spacesuit ready for its inaugural flight and, along with Don Peterson, had performed the first shuttle spacewalk on STS-6. His role as Capcom for our 1990 Hubble deployment mission had given him a good understanding of the telescope and some of its EVA maintenance provisions. There was no arguing that Story was a great pick and very deserving of the assignment, but the news stung nonetheless. Beneath the excitement of my own next mission and musings about my future, some part of me had held out hope that my years of working on Hubble might earn me the plum of leading the first Hubble repair mission. That door had now closed. I chalked it up as one point in favor of a career shift.

*

Sunday, March 22, 1992: two days from liftoff for STS-45. As customary before shuttle flights, we spent the afternoon with our immediate families at the small beach house near the launch pad that Tom Wolfe made famous in *The Right Stuff.* My father and I took a long walk together, just as we had done before each of my prior flights. Though it was never said aloud, we both knew this was our "just in case" walk, the moments, words, and hugs he would cling to if his baby girl never came back. I had told him about the NOAA opportunity when it first came up in January, and we had since spent several long phone calls talking through the pros and cons of different scenarios for my future. "You'd be really happy if this was my last flight, wouldn't you?" I said as we strolled along. Without a word, he pulled from his shirt pocket a tiny Ziploc bag containing a huge pill: the "just in case" tranquilizer he had requested from my flight surgeons. Aerospace engineer that he was, he knew all about the hazards of spaceflight. Supportive parent that he was, he had never burdened me with his anxieties, but just quietly braced himself for the worst instead. After a big, long hug, we continued our stroll. Chalk up another point in favor of the career shift.

*

The axiom "There's no such thing as a bad spaceflight" was proven true again on STS-45. The science was fascinating, our crew very simpatico, and the pace quite manageable. It goes without saying that the views out the window and life in microgravity were as wonderful as ever. A handful of memories remain particularly vivid to this day, two scientific and two human.

Our views of the southern aurora make up the first of my scientific memories. Our trajectory had been designed to meet the ambitious auroral observing goals of the ATLAS-1 science team, and that translated into lots of spectacular views for us. Our night passes were long enough that we could watch the auroral curtain change shape, shift colors, brighten, and fade, even though we were zooming by at five miles per second. The pass that sticks out in my mind consisted of three scenes: In scene 1, I was looking out the window along our line of travel at the shimmering green and red curtain ahead of us. It seemed like a heavy theater curtain, hung in deep folds and moving slowly in response to a gentle breeze. In scene 2, I was above the curtain, looking straight down at it, edge on. For an instant, I thought I could see individual filaments of the Earth's magnetic field, each a luminous green thread created by the cascade of charged particles into the lower atmosphere. Scene 3 started a tiny fraction of a second later, as the aurora again became a shimmering curtain that we were quickly leaving behind.

We did more than simply admire the natural aurora on STS-45; we made some of our own. One of our top-priority payloads was a dose-response experiment that involved firing a pulse of electrons down into the atmosphere and measuring the luminosity it produced with a very sensitive camera. I was down on the shuttle's middeck enjoying dinner with three of my crewmates when the trio working on the flight deck started the first of what were to be many runs of this experiment. I listened with one ear as they went through the checklist leading up to the start command. Suddenly, Dave Leestma shouted "Holy cow—look at *that*!" This was a clear violation of one of the cardinal rules of spaceflight, which states, "There shall be no sentences ending in 'that' (as in 'what the #!*#!! was *that*?!') lest you cause panic in the spacecraft." Food packets were left drifting around the middeck as the four of us zoomed up to the flight deck to see what was going on. We arrived just in time to see the next cycle of the electron beam experiment. An eerie, light blue glow was building up around the top of the coffee-can-shaped instrument mounted out in the payload bay. It seemed to leap away from the shuttle when the "fire" command went out, curving toward the ground under the influence of the Earth's magnetic field. It looked just like the photon torpedoes in a science fiction movie. We were all waiting eagerly for the next pulse when Byron announced that the instrument appeared to have blown a fuse. His guess proved correct, unfortunately. We had fired our last photon torpedo.

George Lucas was the reason we had science fiction movies on our mind during the flight. We had onboard with us the special Oscar that he was to receive at the 64th Academy Awards on March 30.[7] The Academy of Motion Picture Arts and Sciences had persuaded NASA to take it aloft and allow us to make the presentation speech. We were all eager to take it out of its swaddling and pose for our very own Oscar photos before we beamed our presentation down to the live audience in Los Angeles' Dorothy Chandler Pavilion. Mike Foale and I practiced for our crucial role in the presentation. Mike's job was to give the Oscar just the right flick of a wrist so it would do a graceful pirouette as it floated slowly into view of the camera. Mine was to snatch the statue deftly in midtwirl, bring it to a stop at the center of the scene and leave it floating in midair as Charlie finished the presentation speech. The video looked pretty good to us when we replayed it on the shuttle's small television monitors. We hoped it looked suitably spacey to the folks on the ground.[8]

*

Commander Charlie Bolden brought the *Atlantis* in to a smooth landing at the Kennedy Space Center at 6:32 a.m. on April 2, 1992. A few hours later, I was lounging around the conference room inside the crew quarters when someone said there was a call for me. John Knauss was on the line. He was ready to send my name to the White House to begin the nomination process, if I was still interested in the NOAA chief scientist position. How soon could I get to Washington to take the next steps?

Two weeks later I was at the White House, meeting with the head of presidential personnel and filling out nomination papers. Four weeks after that, the White House sent my nomination to the Senate. My career shift was now fully underway, and it felt very good indeed.

*

I saw little point in hanging around Houston as a lame-duck astronaut while waiting for the Senate to confirm me, so I arranged to be detailed to NOAA headquarters. This temporary assignment would let me start learning about my new agency while still keeping one foot anchored firmly in NASA, a prudent hedge against the uncertainties of the Senate confirmation process.

The value of that hedge rose sharply on Tuesday, November 3, 1992, when Governor William J. Clinton of Arkansas soundly defeated incumbent President George H. W. Bush to become the forty-second president of the United States. My nomination to become NOAA chief scientist was instantly null and void. Now what?

By then I had learned enough about the job to know I really wanted it. I had also learned enough about the ways of Washington to believe I could make a strong case for the new administration to renominate me to the post. My plan had two prongs: convince the transition team that I was a good pick on the merits and neutralize any stigma the Bush nomination might carry with the new administration by mustering active support from leading Democratic members of Congress. John Knauss, the outgoing NOAA Administrator, and Barbara H. Franklin, the outgoing Secretary of Commerce, helped with both prongs, vouching for my qualifications and urging the transition team not to turn the earlier nomination into a partisan issue. I began the second prong by visiting with senators and representatives I had met as an astronaut to inform them of my goal and get their advice on how to proceed. This boiled down to keep up the good work, be patient, and add this or that other member of Congress to your visit list.

I continued working with the NOAA transition team and making my rounds on Capitol Hill as November rolled into December and then became January. By the time Ronald H. Brown was sworn in as Secretary of Commerce on January 22, 1993, the White House had received letters supporting my nomination from several members of Congress, most notably Senators Barbara Mikulski and Bill Nelson. I had done all I could to secure the new nomination and now could only wait and hope that the White House got around to NOAA nominations before my detail assignment ended.

It proved to be a very close call. In mid-February, I began lining up a new job at NASA headquarters. I still had no word about the nomination, and the hard end-date of my detail to NOAA was approaching rapidly. I had just returned to the Commerce building after a meeting on this subject at NASA when I happened to bump into Melissa Moss, Secretary Brown's chief of staff. When she asked what was new, I told her about my imminent return to NASA. "But the Secretary wants you to stay as chief scientist," she said, with a look of surprise on her face. I assured her that I wanted that as well, but it was not within my power to make it happen. If he indeed wanted me to stay, the people who had such power needed to do something quickly, before time ran out on me.

The wheels turned very rapidly from that point onward. I met with the Secretary two days later, and the White House resubmitted my nomination to the Senate in late April, along with those of D. James Baker to be NOAA administrator and Douglas K. Hall to be deputy administrator. The Senate Committee on Commerce, Science, and Transportation held a confirmation hearing for all three of us on May 24, and the full Senate confirmed us by unanimous consent on May 28, 1993. I was now officially the Honorable Kathryn D. Sullivan, Chief Scientist, National Oceanic and Atmospheric Administration.

*

Every astronaut longs to be on every flight. Not all will admit to wrestling with pangs of envy whenever they have to watch a mission from the ground. Those pangs hit me hard when *Endeavour* roared off the launch pad on the first Hubble repair mission in the early morning hours of December 2, 1993. The seven lucky astronauts on the STS-61 crew would soon get to see Hubble in orbit—an experience that had eluded me, locked inside the airlock as I was when we deployed the telescope in 1990. Four of them would get the even greater treat of clambering over the telescope, just as I had dreamed of doing during my many long working shifts in the VATA, and putting our tools to the do-or-die test. I loved my new job and knew that coming to NOAA had been the right move for me, but part of me longed to be aboard *Endeavour* nonetheless.

It is impossible to overstate the drama that surrounded this first Hubble maintenance mission. It rivaled that of STS-26, the first shuttle flight after the *Challenger* accident. As at that earlier moment in 1988, the future of the shuttle and its astronauts and even perhaps the fate of NASA itself seemed to be riding on STS-61. Could Hubble's sight really be restored? Would the telescope be written off as a failure if the fix did not work? What would become of the shuttle program, of the space station that was then on the drawing boards, if the mission failed? Such momentous questions were no doubt running through many minds as the shuttle pierced the dawn sky atop a column of fire. As historian Joseph Tatarewicz would later put it, "It seemed they would all come back in the same condition: heroic and vindicated, or disgraced."[9]

Another question joined the list in my mind: How did we do? Had our prelaunch work laid as sound a foundation for maintaining Hubble as we aimed for it to do? What had we missed? This mission was the ultimate test of all the work our original M&R team had done between 1985 and 1990. The next ten days would reveal our final grade.

"Spectacular" was the only word that came to mind whenever I caught bits of the mission on television, but the word fell far short of doing justice to what I was watching. On one level, it all looked very familiar: the telescope I knew so well, the cradle on which it was berthed and the pallet that carried the replacement equipment, the tools and tethers and caddies clustered around the spacewalkers' chests and wafting around the stanchion on the foot restraint attached to the tip of the robotic arm. It was just like the preliminary maintenance simulations we had done in the neutral buoyancy simulator years before, except for the shiny metallic finishes on the maintenance hardware and the Earth gliding by in the background. The familiarity was heartening, for it suggested we had indeed laid a good foundation before we put Hubble into orbit.

Foundations are not ends unto themselves, of course. The structures that get built upon them are what really matter, and the first servicing mission was an elaborate and impressive structure indeed; a veritable cathedral built upon our foundation stone. In the three years since deployment, the Hubble servicing team had expanded the shuttle's capacity from two EVAs per mission to the five slated for STS-61. They had also radically transformed NASA's approach to EVA training, adding a high-fidelity robotic arm to the neutral buoyancy facility, and extending the duration of underwater training sessions by a factor of two, to match the duration of real spacewalks. They had electronically lashed together the neutral buoyancy simulator, the high-fidelity shuttle mission simulators, and the Hubble control center at Goddard, so that the full team could train realistically for maintenance missions, and added virtual reality to the mix of powerful training tools. The rough spacewalk choreography we had developed in the neutral buoyancy water tank had been combined with the movements of the shuttle's robotic arm and command sequences from the Space Telescope Operations Control Center to produce an efficient and elegant orbital dance.

*

Like all grand cathedrals, the Hubble rescue mission had many architects and builders, but everyone directly involved with the mission would agree that Ron Sheffield and the Lockheed M&R team were vital to the success of the hands-on maintenance work. Ron, along with Peter Leung and Brian Woodworth, knew Hubble better than anybody else on the team, having spent hundreds of hours working on it in the VATA and hundreds more coordinating every aspect of all the maintenance procedures with the electrical and mechanical engineers who had built the telescope. The behind-the-scenes work of their teammates Monty Boyd, Rob Lyle, and Roger Phillips had ensured that design changes got done, production schedules were kept, and correct technical information flowed into training manuals and checklists. On the NASA side, Jim Thornton, Sue Rainwater, and Robert Trevino brought to the team a deep and holistic understanding of Hubble EVA operations, formed by their many years of hands-on work with the telescope, with each tool and piece of support equipment, and every page of the crew's EVA checklists and maintenance procedures.

Human nature being what it is, the knowledge and wisdom that these people brought to the new Hubble servicing team was not always welcomed at first. The Goddard team considered themselves to be the world's experts in satellite servicing. They had every right to be proud of having pioneered modular satellite construction, designed the first satellite servicing tools, and led the world's first satellite repair effort on Solar Max in 1984. Unfortunately, that legitimate pride tended to escalate into hubris, which led them to initially ignore much of the work that had already been done, and to dismiss the expertise of the original M&R team.

Just as Bruce and I had feared years earlier, the Goddard team reinvented a number of wheels and duplicated a lot of perfectly good equipment when they took over the on-orbit servicing responsibility. One running joke had it that the real reason Goddard duplicated hardware was so that every Hubble tool would be "Cepi blue," the particular shade preferred by group leader

Frank Cepollina, like the module service tool that was used on the Solar Max repair mission in 1984 (figure 9.2). Many of the tools that Marshall had designed for handling and protecting units during maintenance—handles, instrument lightshades, and the like—were discarded. The example I found particularly absurd was making a Goddard Go/No Go gauge to replace the one that Lockheed had built and tested on every single fastener before we took Hubble to orbit.

Goddard's initial disregard for the Lockheed team's expertise soon wore off. Near the end of 1991, they turned all responsibility

FIGURE 9.2

The "Cepi blue" module service tool used on the Solar Maximum satellite rescue mission in 1984. Source: NASA.

for neutral buoyancy mockups and testing over to Sheffield and his crew. Soon after the rescue mission landed, Cepi began lobbying Ron, Peter, and Brian to relocate permanently to Maryland. For once, his considerable powers of persuasion failed; they stayed in Sunnyvale.

The four astronauts who would actually make the repairs also pushed back on the inherited work at first. Story Musgrave, Jeff Hoffman, Tom Akers, and Kathy Thornton all had a spacewalk under their belt (Akers had two) and each naturally and rightly approached the Hubble tasks from the vantage point of their own prior experience. Story in particular had strong views about EVA techniques that sometimes clashed with the baseline plans developed in the 1985–1990 timeframe. He believed, for example, that many tasks could be done free-floating, without being anchored in a portable foot restraint. The only place to resolve questions like this and turn baseline plans into detailed, efficient flight procedures was the neutral buoyancy tank. This was also where the crew and EVA team supporting them would fine-tune the sequence of tasks to maximize efficiency and develop a safe working tempo (figure 9.3). Move too fast, and the crew risked a damaging collision; too slow, and they risked delicate hardware getting too cold.

I only spotted one minor instance of blindness or "not invented here" affecting the EVA operations as I followed the mission on television. On the final spacewalk, Story Musgrave and Jeff Hoffman had opened two of the large doors on the bottom end, or "aft shroud," of the telescope to gain access to the star trackers and rate-sensing units that are crucial to Hubble's guidance system.[10] These swing open like enormous kitchen cabinet doors once they are unlatched. Closing them involves a bit of a trick, however. Hubble's three star-trackers are mounted in a cluster behind these doors. This forced the telescope's design engineers to make one door twice as wide as the other and to cut

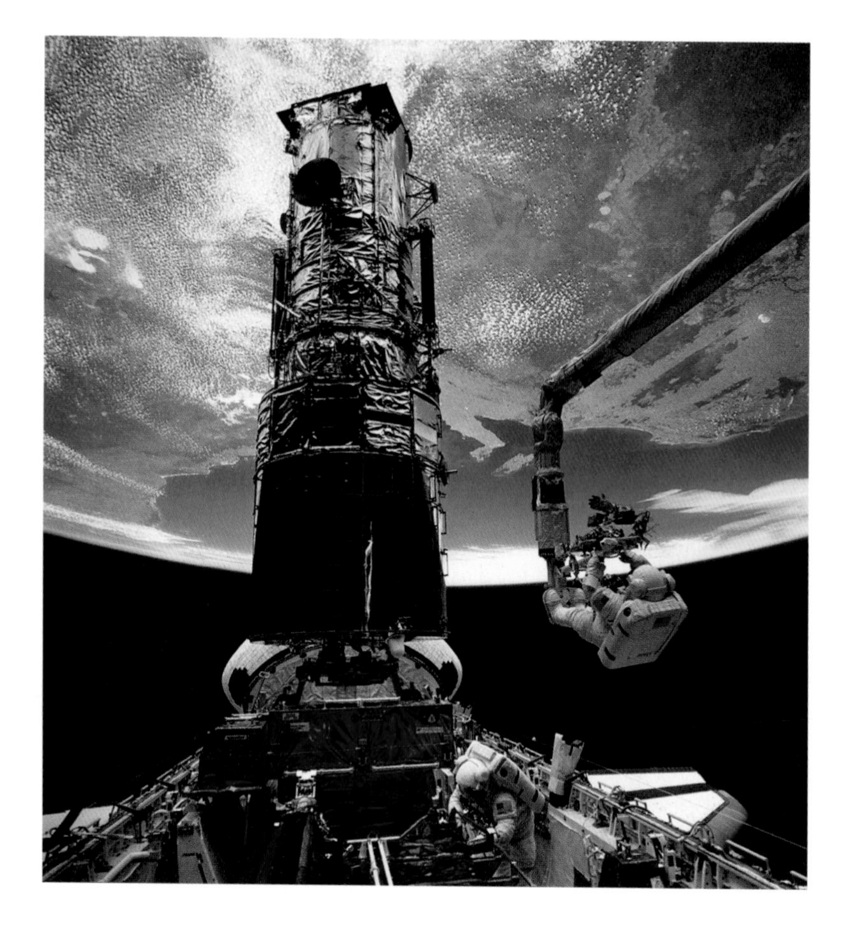

FIGURE 9.3

Jeff Hoffman riding on the end of the shuttle's robotic arm during the first Hubble servicing mission, with Story Musgrave working below him in the cargo bay. Source: NASA.

three large openings so that light could reach the star trackers (figure 9.4). Our prelaunch work in the VATA had taught us exactly where to apply pressure on the larger door to make it seat properly with its mate. This technique had been photographed and documented thoroughly, including in the hefty volume of detailed

FIGURE 9.4

Lower, “aft shroud” section of Hubble, showing the three star-tracker ports (two black ovals and black circle just below the white panel of the Wide Field and Planetary Camera 2). Source: NASA.

photographs that Sheffield's crew produced for the EVA support teams at Goddard and Johnson. Reading about something and really knowing how to do it are two different things, however. Training is supposed to close that gap, but failed to in this case. The underwater mockups were not highly accurate replicas of the aft shroud structure, so the quirk did not occur in the neutral buoyancy simulator—the one and only place where the crew worked with the aft shroud doors. Ron and Bruce had both lobbied hard for Goddard to build an exact replica of the aft shroud structure, so that crews could develop the touch and feel of working in this sensitive area of the telescope, but their pleas had been met by deaf ears and firmly closed pocketbooks.

The report from *Endeavour* that the doors would not close caused quite a kerfuffle in mission control at first. Happily, Sheffield and his team were staffing one of the engineering support consoles in what's known as the "back room" of mission control, and Bruce McCandless was present to coach the flight control team and Capcom on what needed to be done. The confusion was resolved in short order and, once Jeff Hoffman pushed on the proper spot, the doors closed without further difficulty.

✷

All of NASA seemed to breathe an audible sigh of relief when the crew of *Endeavour* set Hubble free again on December 9, 1993. The astronomical community heaved an equally deep sigh nine days later, as the data came in that confirmed the corrective optics installed by the crew had indeed restored Hubble's sight (figure 9.5).[11] Senator Barbara Mikulski, who had blasted Hubble as a "techno-turkey" when the mirror flaw was discovered, now proclaimed, "The trouble with Hubble is over." The twinges of envy I had felt at launch were long gone, replaced by joy and the kind of

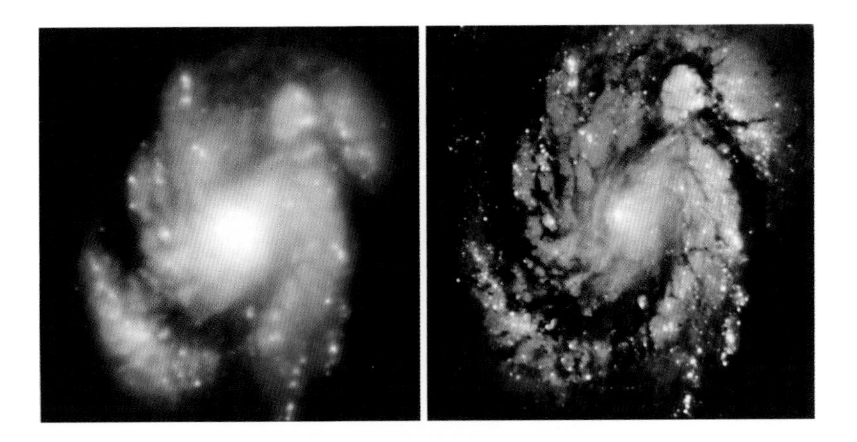

FIGURE 9.5

Hubble's view of the M100 Galactic Nucleus before the optical correction and after. Source: NASA.

warm pride that alumni feel when their old college team wins a national championship.

The first of many honors that the servicing mission team would receive was awarded in the middle of the night on December 13, 1993. *Endeavour* touched down on the Kennedy Space Center runway just before 2:30 a.m. Eastern Standard Time. About an hour later, as the crew of seven astronauts was disembarking, shuttle mission control in Houston passed responsibility for the spacecraft back to the ground operations team at Kennedy. Flight controllers from each of the three round-the-clock shifts that had supported the mission crowded into the main control room to witness the traditional ceremony that marks the end of a successful mission at Johnson.

Hanging the plaque—adding the emblem of the just-completed flight to the array of mission plaques lining the walls of the control room—is the highlight of this celebration (figure 9.6).

FIGURE 9.6

The STS-61 mission emblem that Jim Thornton hung in shuttle mission control after the successful conclusion of the first Hubble servicing mission. Source: NASA.

The flight director awards the honor of climbing the ladder to hang the plaque to the team whose work was most essential to the mission's success. There is often some suspense around who will get this coveted honor, but there wasn't any that morning. Everyone knew it belonged to the EVA team.

Jim Thornton, the EVA lead at Johnson, would climb the ladder alone, but he would be climbing for everyone on the satellite servicing team at Goddard and Sheffield's crew from Lockheed, as well as the large supporting cast of tool designers, suit technicians, machinists, and maintainability engineers whose work reached back fully twenty years.

*

Thinking ahead about honors that were yet to come, I sat down at my desk the next morning to pen a letter to Goddard Center director Dr. John Klineberg, with courtesy copies to Joe Rothenberg, head of flight projects at Goddard, servicing mission flight director Milt Heflin at Johnson, and mission commander Dick Covey. I knew that a hands-on engineering group like Ron's would normally not be singled out to receive any of the most prestigious awards that were sure to come. This was a simple matter of scale. The servicing mission team was so large and involved so many organizations that nominators would have to choose a few high-level leaders to symbolize the entire team. I understood the logic of that, but felt strongly that an exception should be made for Ron and his team. Their work before the deployment mission had been absolutely essential to turning the concept of Hubble maintenance into a real and effective capability. Their knowledge of both the telescope and the workings of NASA had been a stabilizing force through the bureaucratic transfer of servicing responsibility from Marshall to Goddard. Finally, the presence of Ron and his team from the inception of maintenance development through first servicing mission operations had provided invaluable continuity of knowledge about the fine details of each tool and maintenance task. In my mind, they were the glue that held Hubble maintenance operations together, and they deserved to be recognized as such.

Fortunately, I had an ace up my sleeve: the elegant official stationery of the Office of the Chief Scientist, National Oceanic and Atmospheric Administration. The combination of my name and Hubble history with my new official station in life guaranteed that my letter would reach all the addressees personally and very likely end up on the desk of the NASA administrator as well. I opened it with hearty congratulations on the very successful mission and then offered the blunt suggestion that a particular individual on the Lockheed M&R team deserved NASA's highest civilian award, the Distinguished Public Service Award, an honor that would normally be given to people several levels higher than Ron in an organization. Then I laid out the most forceful and eloquent arguments I could muster in favor of this, summarizing the vast scope of the work, the exacting detail it demanded, and Ron's personal dedication and leadership. In closing, I noted that Ron Sheffield exemplified "the superlative, dedicated technical experts who quietly and selflessly undergird the success of so many NASA evolutions" and urged that they recognize the vital behind-the-scenes work of the M&R team. I signed it with a flourish and sent my letter on its way on December 15, 1993.

*

The Collier Trophy is to American aviation and spaceflight what the Stanley Cup is to hockey or the Oscar is to filmdom: the highest and most coveted of all awards.[12] Established in 1911, it is presented annually by the National Aeronautic Association "for the greatest achievement in aeronautics or astronautics in America." In March 1994, the association announced that the 1993 award would go to the Hubble Space Telescope Recovery Team "for outstanding leadership, intrepidity, and the renewal of public faith in America's space program by the successful orbital recovery and repair of the Hubble Space Telescope."[13] The eleven named recipients—seven astronauts and four senior-level NASA managers—represented a team of more than 1,200 men and women who were directly involved in the mission.

Tickets to the Collier dinner at the Smithsonian's National Air and Space Museum in Washington, DC, are very hard to come by. This black-tie affair is a premier event on the aerospace world's annual calendar, attracting the Who's Who of the field from industry and government. Guests mingle before dinner among such legendary craft as the Wright Flyer and John Glenn's Friendship 7 Mercury capsule. The Spirit of St. Louis hangs above the honorees as they accept the trophy. Ron and Linda Sheffield joined the A-listers on May 6, 1994, as the invited guests of *Endeavour*'s commander, Dick Covey. Ron was the only Lockheed employee in attendance and, as far as he could tell, the only non-NASA member of the Hubble repair team in the room.[14]

*

The auditorium at NASA headquarters was buzzing with happy energy as the crowd gathered on May 19 for the agency's annual awards ceremony. I had hooted with joy when the NASA administrator's invitation to attend the event hit my desk a few weeks earlier, knowing it probably meant my letter to Klineberg had produced the intended result. An email to Ron confirmed that he was, indeed, to receive NASA's Distinguished Public Service medal, alongside his boss, Tom Dougherty. Making my way to my seat in the auditorium that day was like crossing the room at a school reunion, with former colleagues stopping me every few feet to hug or shake hands, and swap a memory or two. I finally spotted my astronaut colleagues and hurried over to congratulate them. "You're not the main reason I'm here, you know," I said as I hugged Dick Covey. "We know," he replied, pointing to his left. "He's right over there." Sheffield and his wife were beaming like a pair of kids in a candy store over the delightful surprise of receiving such an unexpected and prestigious award.

✷

Four other shuttle crews would visit Hubble over the next sixteen years. Each mission built upon the foundation it inherited, which reached all the way back to Hubble's design engineers. Every servicing mission team improved upon the existing methods and invented new devices to tackle ever more complex repairs, but all relied heavily on the tools and equipment produced by the original M&R team. By the fourth mission, the Hubble repair teams were doing things that we never would have contemplated back in 1990, such as taking the lid off a delicate scientific instrument mounted just a few feet below the primary mirror in order to change individual circuit boards. That is like precision microsurgery compared to the tasks we had prepared for, and it required some impressive innovations.[15]

Over the course of the five maintenance missions, sixteen spacewalkers would spend a total of 165.8 hours—just six minutes shy of seven full days—riding the shuttle's robotic arm or clambering around the telescope. Thanks to these missions, the telescope today is a far better instrument than the one we first deployed on April 25, 1990. A threefold increase in sensitivity allows it to see deeper into the universe. Its sensors cover more wavelengths of light and gather greater volumes of data. As a machine, it is also more reliable and more efficient, with fewer moving parts to fail and solar arrays that produce 20 percent more electricity despite being one-third smaller than the original set. Essentially all that remains of the Hubble we carried aloft in 1990 are the two mirrors and the metering truss that holds them, the yellow EVA handrails and foot restraint sockets, and the shiny silver outer skin that makes this magnificent flying machine so easy to spot when it passes overhead at twilight.

✷

That silver skin sparkles much less than it did on the day I first saw Hubble back in April 1985. Much of the decay is due simply to the harsh environment of space. All satellites "weather" as they are bombarded constantly by micrometeoroids, bits of space debris, intense radiation, and the charged particles in the solar wind. Hubble is alone among satellites in being affected by another weathering force: human contact. As they went about their vital tasks, the spacewalking astronauts who maintained the telescope left their handprints on its outer skin (figure 9.7).

FIGURE 9.7

Scuff marks on Hubble's outer skin left by the gloved hands of spacesuited maintainers. Source: NASA.

These visible handprints are like the tip of an iceberg, dramatic hints of a larger mass that is out of sight. To me they symbolize the countless earthbound hands that designed maintainability into the telescope, built the tools and equipment needed to make on-orbit servicing a reality, trained the flight crews, and worked each mission as tirelessly as the astronauts themselves. Each of these unsung people can rightly claim to also have left their handprints on Hubble.

NOTES

CHAPTER 1

1. As the telescope entered its twenty-fifth year of operations in 2015, NASA reported that over 4,000 astronomers from around the world had used Hubble. The telescope had made over 1 million unique observations, produced over 100 terabytes of data and resulted in more than 11,000 scientific papers. As of 2018, 19,799 scientists have been listed as authors of papers based on Hubble data, and these articles have been cited 811,117 times. Over 12,000 users from all 50 states and 85 countries retrieve 6 to 15 terabytes every month from the 160 terabyte Hubble archive. For more information see: https://www.nasa.gov/content/goddard/a-look-at-the-numbers-as-nasas-hubble-space-telescope-enters-its-25th-year; https://archive.stsci.edu/hst/bibliography/pubstat.html; https://hst-docs.stsci.edu/display/HSP/Hubble+Space+Telescope+Science+Policies+Group+and+Peer+Review+Information.

2. Lyman S. Spitzer Jr. (1914–1997). Lyman Spitzer, *The Astronomical Advantages of an Extra-terrestrial Observatory*, Project RAND Report, Douglas Aircraft Company, September 1, 1946. Hermann Oberth had mentioned the possibility of using a rocket to place a telescope into space in his 1932 *Die Raketen zu den Planeträumen*, but the engineering capacity of his times was not up to the task.

3. The Large Space Telescope would not be named for astronomer Edwin P. Hubble until 1983.

4. NASA has more acronyms than some languages have words. I have avoided them as much as possible, with the notable exceptions of "EVA" and "M&R" (explained below).

CHAPTER 2

1. According to the Survey of Earned Doctorates published by the US National Center for Science and Engineering Statistics, women earned 3,512 of the 17,695 doctoral degrees awarded in the sciences and engineering in 1978. http://www.norc.org/PDFs/publications/SED_Sum_Rpt_1998.pdf.

2. Comparing space to the sea is a common metaphor. In his famous "We go to the moon" speech, John F. Kennedy said, "We set sail on this new sea because there is new knowledge to be gained …" (https://er.jsc.nasa.gov/seh/ricetalk.htm). See also David Meerman Scott and Ricahrd Jurek, *Marketing the Moon: The Selling of the Apollo Lunar Program* (Cambridge, MA: MIT Press, 2014).

3. Renamed the Lamont-Doherty Earth Observatory in 1993.

4. In addition to George Abbey, the interview panel included Joseph D. Atkinson Jr., PhD; astronaut Vance D. Brand; astronaut Edward Gibson, PhD; Carolyn Huntoon, PhD; astronaut Joseph P. Kerwin, MD; Jack R. Lister; Glynn S. Lunney; astronaut Robert A. Parker (who had not yet flown in space); Robert O. Piland; Martin L. Raines; Duane L. Ross; James H. Trainor, PhD; and astronaut John W. Young.

5. Sally and I knew from the press coverage surrounding our selection that we had grown up near each other, in the San Fernando Valley section of Los Angeles. Our very first conversation turned quickly to comparing street addresses and school history to see how close we really had been. To our great surprise, we went to the same school in first grade, Hayvenhurst Elementary School in Van Nuys. Whether we had been in the same classroom or ever crossed paths on the playground is another question, and one that neither of us could answer.

CHAPTER 3

1. Ratner's commentary for both the scrubbed launch attempt on April 10 and the successful launch on April 12, 1981, can be found at https://www.youtube.com/watch?v=gz1r99UCAW0.

2. Inclination is the angle between the plane of an orbit and the equatorial plane. The inclination angle translates into the maximum latitude of a spacecraft's ground trajectory or, from an astronaut's point of view, the chance to see much more of the Earth and possibly the aurora. With few exceptions, the space shuttle flew in either 28.5 degree or 57 degree orbits. The lower inclination allowed the shuttle to launch due east out of Kennedy Space Center, taking maximum advantage of the Earth's rotational velocity. The 57-degree inclination allowed greater ground coverage for earth science experiments and overflight of partner space facilities in Japan, Canada, Europe, and, nowadays, Russia.

3. The proper usage of "in orbit" versus "on orbit" is a matter of long-standing debate. NASA commonly uses "on-orbit" (hyphenated) rather than "in orbit" when referring to where an activity occurs, as in "on-orbit refueling" or an "on-orbit status report," whereas major news organizations almost always use "in orbit." I have used "on-orbit" where needed to align with titles or terminology used in the projects I describe and applied the more common press usage elsewhere. Robinson Meyer provides a longer and quite amusing discussion of this arcane debate in his article, "Grammar in Space: Are Satellites 'In Orbit' or 'On Orbit'?" published in the November 9, 2014, issue of the *Atlantic*.

4. Of course, there were other explanations for the new system. According to NASA, "Some have suggested this system was created solely to avoid then-NASA Administrator James Beggs' triskaidekaphobia, or fear of the number 13. ... In reality, NASA anticipated a much greater frequency of launches as the Space Shuttle system moved into full operations. Plans called for up to 50 launches per year with Vandenberg AFB [in California] launching Shuttles into a polar orbit, while Kennedy launched them into an equatorial orbit. On top of that, NASA was planning the payloads and launch dates for the Shuttles years in advance. Delays and cancelations threatened to push the Shuttles out of sequential order anyway, so the new system was designed to prevent this confusion. Unfortunately, it was not very successful in this department. Shuttles were still launched out of order" (https://www.nasa.gov/feature/behind-the-space-shuttle-mission-numbering-system). The new system did nothing to change the schedule volatility, but simply masked it under a more confusing mix of numbers and letters.

5. Abrahamson memo, 1983, Space Telescope History Project collection, National Air and Space Museum.

6. Valerie Neal, *Spaceflight in the Shuttle Era and Beyond: Redefining Humanity's Purpose in Space* (New Haven: Yale University Press, 2017).

CHAPTER 4

1. Earth's atmosphere blocks light in the infrared and ultraviolet wavelength regions that are important to optical astronomers.

2. David H. DeVorkin, *Science with a Vengeance: How the Military Created the U.S. Space Sciences after World War II* (New York: Springer, 1992).

3. Lyman Spitzer Jr., *The Astronomical Advantages of an Extra-terrestrial Observatory*, Project RAND Report, Douglas Aircraft Company, September 1, 1946.

4. Spitzer, *Astronomical Advantages of an Extra-terrestrial Observatory*, 131–132.

5. Eric Chaisson, *The Hubble Wars; Astrophysics Meets Astropolitics in the Two-Billion-Dollar Struggle Over the Hubble Space Telescope* (New York: Harper Collins, 1994).

6. Chaisson, *Hubble Wars*.

7. Comprehensive accounts of the telescope's scientific and political history can be found in Chaisson, *Hubble Wars*; Roger D. Launius and David DeVorkin, *Hubble's Legacy: Reflections of Those Who Dreamed It, Built It, and Observed the Universe with It* (Washington, DC: Smithsonian Institution Scholarly Press, 2015); David J. Shayler and David M. Harland, The HST: From Concept to Success (New York: Springer Praxis, 2016); and Robert W. Smith, with Paul A. Hanle, Robert H. Kargon, and Joseph N. Tatarewicz, *The Space Telescope: A Study of NASA, Science, Technology and Politics* (Cambridge: Cambridge University Press, 1989). James J. Mattice, in *Hubble Space Telescope Systems Engineering Case Study* (Dayton, OH: Center for Systems Engineering, Air Force Institute of Technology, Wright-Patterson AFB, 2005), presents a systems engineering perspective on the development.

8. Explorer 1 was built and operated by the Jet Propulsion Laboratory, under the direction of Dr. William H. Pickering. Wernher von Braun designed the Jupiter-C rocket that carried it aloft. Dr. James Van Allen from Iowa State University built the scientific payload, a cosmic-ray detector. Data from Explorer 1 led Van Allen to theorize that the Earth was surrounded by belts of charged particles trapped by the magnetic field. When another satellite confirmed the existence of these belts two months later, they were named the Van Allen Belts in honor of their discoverer. For more information see https://www.nasa.gov/mission_pages/rbsp/mission/fun-facts.html or https://www.nasa.gov/content/goddard/van-allen-probes-mark-first-anniversary/.

9. Quoted in Lyman Spitzer Jr., "History of the Space Telescope," *Quarterly Journal of the Royal Astronomical Society* 20 (1979): 31.

10. Grumman Aircraft Engineering Corporation, "The OAO in Association with a Manned Space Station," NASA contract NASW-1143, 1965, Space Telescope History Project, Acc. 1999-0035, National Air and Space Museum, Smithsonian Institution, Box 31; National Research Council, *Space Research: Directions for the Future* (Washington, DC: The National Academies Press, 1966).

11. Boilerplate models of the capsule had flown several times in 1965, but the first flight of a fully functional Apollo capsule did not occur until Apollo 4 (AS-501) in November 1967. See Roger E. Bilstein, *Stages to Saturn: A Technological History of the Apollo/Saturn Launch Vehicles*, NASA SP-4206 (Washington DC: National Aeronautics and Space Administration, 1996), and https://nssdc.gsfc.nasa.gov/planetary/lunar/apollo.html.

12. Quoted in Spitzer, "History of the Space Telescope," 32.

13. Quoted in Spitzer, 32.

14. The first years of Large Space Telescope (LST) development do not fit neatly into the project development stages NASA uses today. According to Dr. Nancy Roman, LST Phase A began in 1965. Dr. Max Nein from Marshall's Program Development Office has stated that there was no true Phase A, but just a bevy of independent conceptual studies commissioned by both Goddard and Marshall that eventually prompted NASA to designate a lead center and launch a formal Phase B design definition effort (NGR and MEN Oral Histories, Space Telescope History Project).

15. John Logsdon, *After Apollo? Richard Nixon and the American Space Program* (New York: Palgrave Macmillan, 2015).

16 Murphy & Darwin (1973), 4-20-17; Grumman Aircraft Corportation (1970).

17. J. T. Murphy and C. R. Darwin, "Lower Payload Costs through Refurbishment and Module Replacement," *Astronautics and Aeronautics* 11 (May 1973): 40–47.

18. The following is based on JRO Oral History, 1-8-87, STHP and discussions with the author.

19. A comprehensive overview of the myriad technical challenges this posed can be found in National Aeronautics and Space Administration, *The Space Telescope*, SP-392 (Washington, DC: National Aeronautics and Space Administration, 1976).

20. Fifteen-month Phase B study contracts were awarded to Boeing, the Lockheed Missiles and Space Company and Martin Marietta Corporation in December 1974 (NASA, *Space Telescope*).

21. The scope of this "full" capability would change multiple times before Hubble launched.

22. Marshall Space Flight Center, "ST Support Systems Module Request for Proposals. NAS8-1-6-PP-00596Q," 1977, Space Telescope History Project, Acc. 1999-0035, National Air and Space Museum, Smithsonian Institution, Box 32.

23. Specifically, a 1/4-28 fastener. 1/4-inch fasteners have a 7/16-inch head on them. The purpose of double-height head was to ensure that the wrench would stay firmly seated even in cases where the astronaut's work position was not ideal.

24. Lockheed Missiles and Space Corporation, "Final Design Review Briefing, Support Systems Module Phase B. DR-MA-03, DPD 449," February 11, 1976.

CHAPTER 5

1. McCandless oral history interview with Joseph N. Tatarewicz, January 8–9, 1986.

2. Withey was employed by ILC Dover, the company that then had the contract to design, produce, and manage the tools and equipment shuttle crews would use both inside the spacecraft and during EVAs. ILC Dover was later acquired by Oceaneering International, which continues to provide these services.

3. Notable tenants on the site today include Yahoo, Juniper Networks, and Amazon Lab 126.

4. In 1986, AFS Sunnyvale was renamed Onizuka Air Force Station in honor of Lt. Col. Ellison S. Onizuka, one of the astronauts who died in the *Challenger* accident. The Defense Department closed the base in 2005, and began full-scale demolition of the buildings in 2014. The iconic Blue Cube no longer exists.

5. The Skunk Works originated in 1943, when Lockheed engineer Kelly Johnson assembled a hand-picked group of forty engineers and manufacturing people to swiftly and secretly build a jet fighter (the XP-80) for the US Army Air Force. The group first worked in a rented circus tent next to a manufacturing plant that produced a horrible odor. The founding team adopted the "Skunk Works" moniker from Al Capp's comic strip *L'il Abner*, which featured a running joke about a smelly and mysterious place deepin the forest called the "Skonk Works." The moniker stuck, and is now a registered trademark of the Lockheed Martin Corporation. Notable aircraft designed and produced by the Skunk Works include the P-38 Lightning, P-80 Shooting Star, U-2 Dragon Lady, SR-71 Blackbird, F-117 Nighthawk, and F-22 Raptor (see https://www.lockheedmartin.com/en-us/who-we-are/business-areas/aeronautics/skunkworks/skunk-works-origin-story.html).

6. See Robert W. Smith, with Paul A. Hanle, Robert H. Kargon, and Joseph N. Tatarewicz, *The Space Telescope: A Study of NASA, Science, Technology and Politics* (Cambridge: Cambridge University Press, 1989), for a comprehensive history of the program's budget and schedule problems.

7. Costa interview with the author, October 2017.

8. Costa interview with Joseph N. Tatarewicz, March 31, 1989.

9. Telephone interview and email exchanges with the author, 2018.

10. Mika McKinnon summarized this incident in a March 2015 article for Gizmodo. During his twelve-minute spacewalk in 1965, Alexey Arkhipovich Leonov's spacesuit bloated and stiffened so much that he could not get back into the airlock of the Voskhod 2 capsule. Opening a valve to release oxygen and soften the suit allowed him to squeeze into the airlock but nearly gave him a bout of the bends, or decompression sickness. Then, when he was safely back inside, jettisoning the inflatable airlock put the capsule into a spin.

11. The original Wide Field/Planetary Camera (WF/PC) was replaced with WFPC2 on the first servicing mission in 1993. WFPC2 included a set of mirrors that compensated for Hubble's flawed primary mirror. It was removed from Hubble in 2009 and is now on display at the Smithsonian's National Air and Space Museum in Washington, DC. Many elements of the original WF/PC were refurbished and used to build Wide Field Camera 3, which replaced WFPC2 in 2009.

CHAPTER 6

1. June would later invite me to join the fledgling Challenger Center for Space Science Education as Vice President for Education and to design the Center's educational program. The design workshop I planned and led at Tucson's SunSpace Ranch in June 1987 produced the Challenger Learning Center concept and the commitment of the Houston Museum of Natural Science to be the inaugural site. The Challenger Learning Center network now spans thirty-one states, plus Canada, South Korea, and the United Kingdom.

2. William P. Rogers et al., *Report of the Presidential Commission on the Space Shuttle Challenger Accident* (Washington, DC: National Aeronautics and Space Administration, 1986); Richard P. Feynman, "Mr. Feynman Goes to Washington," *Engineering and Science* 51, no. 1 (1987): 6–22; Thomas F. Gieryn and Anne E. Figert, "Ingredients for a Theory of Science in Society: O-Rings, Ice, Water, C-Clamp, Richard Feynman and the New York Times" in *Theories of Science in Society*, ed. Susan Cozzens and Thomas F. Gieryn (Bloomington: Indiana University Press, 1991), 67–97.

3. I have been unable to confirm who participated, but it was most likely one or both of Jim Thornton and Sue Boyd Rainwater.

4. Testing tools on Hubble itself was the only option at this time. Several years later, Goddard would build a replica of Hubble's equipment section that was of high enough fidelity to test EVA tools and give astronauts a realistic sense of what it would feel like to operate fasteners and connectors on-orbit.

CHAPTER 7

1. https://history.nasa.gov/rogersrep/genindex.htm.

2. Working on this book spurred me to try to reconnect with the artist and ask him about his gift, which now hangs in the foyer of my home. The inscription penciled onto the margin had long since become invisible, taking with it his name. Counting on the power of networks, I circulated a photograph of the piece among the members of the Association of Space Explorers, the professional and educational society of astronauts and cosmonauts. This yielded Anatoly Ivanovich Veselov's name plus the phone number for his home in a town outside of Moscow. I then asked an Armenian-born, Russian-speaking colleague from NOAA, Yana Gevorgyan, to call the Veselov home and ask if he would be willing to talk or exchange letters with me about our encounter during the Sputnik festivities and his etching. Unfortunately, his wife, Galina Ivanova, informed Yana that he was suffering from advanced Parkinson's disease and could no longer hear or speak. That ended my hope of finally learning the story behind my piece of art and, I thought, of getting permission to reproduce the piece in this book. As I tried to rewrite the story so that it could stand on its own, without a picture, I kept thinking back to Yana's description of Galina Ivanova's sadness that her husband and his works seemed to have been forgotten. There had to be a way to solve the permissions problem and get this unique piece of art, which I believe has never been seen in public, into print for at least a small slice of the world to appreciate. But how to overcome the language barrier and get my request to her reliably? The answer turned out to be a childhood friend of Yana's, now living in Moscow, who volunteered to serve as courier. The permissions form, a thank you note and some cash to buy flowers or bon bons for Galina Ivanova went by express mail to Maria in Moscow in mid-November 2018. Sadly, Anatoly Ivanovich passed away while the package was en route. Galina Ivanova very graciously agreed to receive Maria a week later and to allow me to reproduce her husband's work in this volume. I learned its title, "Assemblage of Orbital Station," thanks to a brochure she sent along with the signed letter of permission.

3. https://www.youtube.com/watch?v=S7WJtQYU8i4. Lyrics reprinted with permission of Michael J. Cahill.

4. According to the *Cambridge Dictionary*, "not invented here" syndrome is "the idea that a product, system, etc. that was developed somewhere else cannot be as good as one that a company, etc. develops itself" (https://dictionary.cambridge.org/dictionary/english/not-invented-here-syndrome).

5. Kapton is the brand name of a polyamide material produced by DuPont. It is widely used in spacecraft design because it is a good insulator, is stable across a wide range of temperatures, and does not out-gas much in the vacuum of space.

6. Michael J. Massimino, *Spaceman: An Astronaut's Unlikely Journey to Unlock the Secrets of the Universe* (New York: Crown Archetype, 2016), 185.

7. For more, see Joseph N. Tatarewicz, "The Hubble Space Telescope Servicing Mission," in *From Engineering Science to Big Science: The NACA and NASA Collier Trophy Research Project Winners*, ed. Pamela E. Mack, NASA SP-4219, 365–396 (Washington, DC: National Aeronautics and Space Administration, 1998).

8. Hubble is outside the atmosphere as we usually think of it, but not entirely outside the gaseous envelope of Earth. It flies in a layer of Earth's upper atmosphere known as the exosphere where the pressure is just 0.0001 percent of the atmospheric pressure at sea level.

9. D. G. King-Hele and Eileen Quinn, "The Variation of Upper-Atmosphere Density between Sunspot Maximum (1957–1958) and Minimum (1964)," *Journal of Atmospheric and Terrestrial Physics* 27, no. 2 (February 1965): 197–209; https://www.swpc.noaa.gov/impacts/satellite-drag.

CHAPTER 8

1. One reason we were an all-veteran crew was to lower the odds of this and allow Flight Day 2 EVA. About half of first-time flyers get space motion sickness. Few have it again on subsequent flights.

2. Hubble's magnitude is approximately +1, equivalent to Spica or Acrux, the brightest star in the Southern Hemisphere. See http://sen.com/news/hunting-hubble-how-to-spot-the-space-telescope.

CHAPTER 9

1. Eric Chaisson, *The Hubble Wars; Astrophysics Meets Astropolitics in the Two-Billion-Dollar Struggle Over the Hubble Space Telescope* (New York: Harper Collins, 1994), 175.

2. The climate of outrage and gloom obscured the fact that Hubble was still quite capable of producing unique and important scientific results, even with the flawed mirror. As Chris Gainor notes in chapter 3 of his *Hubble Space Telescope Operating History* (forthcoming), its ultraviolet observations were still better than anything available before. The spectrographic and photometric instruments were not affected as strongly as the imaging instruments. In spite of the mirror issue, an early image of the Large Magellanic Cloud taken by the Wide Field/Planetary Camera showed

sixty stars clearly, whereas the best ground-based images showed only eight clear ones and hinted at perhaps as many as twenty-seven. Scientists from the Space Telescope Science Institute in Baltimore testified to Congress that 13 percent of the observations planned for the first observing cycle were unaffected by the flaw, and 56 percent were still viable in spite of it.

3. Lampton would later have to withdraw for medical reasons. Belgian engineer and physicist Dirk Frimout would take his place.

4. Excluded initially from the group, Bruce called his Naval Academy classmate "Johnny," a.k.a. Senator John McCain, who threatened to withhold the project's funding unless he was added to the team.

5. For a more complete description of the mirror problem and how the solution emerged, see Chaisson, *Hubble Wars*; Robert W. Smith, with Paul A. Hanle, Robert H. Kargon, and Joseph N. Tatarewicz, *The Space Telescope: A Study of NASA, Science, Technology and Politics* (Cambridge: Cambridge University Press, 1989); and Joseph N. Tatarewicz, "The Hubble Space Telescope Servicing Mission," in *From Engineering Science to Big Science: The NACA and NASA Collier Trophy Research Project Winners*, ed. Pamela E. Mack, NASA SP-4219 (Washington, DC: National Aeronautics and Space Administration, 1998), 365–396.

6. NOAA is a semi-independent unit within the Department of Commerce, making up over half of the department's budget and a quarter of its personnel. The agency's mission is to understand how our planet works and turn that understanding into reliable and useful information such as weather forecasts, nautical charts, and fish stock assessments. This requires both a sound research foundation and the ability to collect vast amounts of data about conditions in the ocean and atmosphere. NOAA's satellites, ships, and aircraft collect the needed data; the agency's scientists and academic partners produce maps, forecasts, and assessments and conduct research in all my favorite disciplines.

7. Lucas received the Irving Thalberg Memorial Award, which is presented to "creative producers whose bodies of work reflect a consistently high quality of motion picture production." See http://www.oscars.org/governors/thalberg.

8. Video of the full presentation can be found at https://www.youtube.com/watch?v=USJNgbfnpQE.

9. Joseph N. Tatarewicz, "The Hubble Space Telescope Servicing Mission," in *From Engineering Science to Big Science: The NACA and NASA Collier Trophy Research Project Winners*, ed. Pamela E. Mack, NASA SP-4219, 365–396 (Washington, DC: National Aeronautics and Space Administration, 1998).

10. A star tracker is an optical device that measures the position of stars relative to the body of a spacecraft. The spacecraft's orientation in space can be calculated by comparing three measured star positions with the known absolute position of the stars provided by astronomical star catalogs.

11. The mirrors on the Corrective Optics Space Telescope Axial Replacement (COSTAR), which replaced the High-Speed Photometer in the telescope's aft compartment, corrected the light that fed into the Fine Guidance Sensors and three remaining science instruments (High-Resolution Spectrograph, Faint Object Camera, and Faint Object Spectrograph). The optical correction was built into the replacement Wide Field and Planetary Camera (WFPC2) during production.

12. NASA Headquarters Press Release 94-71. See also Martin Cohen recollection of WFPC-2 and STS-61 published at http://www.company7.com/c7news/nasa_sts61.html.

13. Tatarewicz, "Hubble Space Telescope Servicing Mission." The HST Recovery Team named in the citation was composed of Joseph Rothenberg, previously Associate Director of Flight Projects, Goddard Space Flight Center, Greenbelt, MD; Randy Brinkley, STS-61 Mission Director, Johnson Space Center (JSC), Houston; James M. "Milt" Heflin, Jr., STS-61 Lead Flight Director, JSC; Brewster H. Shaw, Jr., Director, Space Shuttle Operations, NASA Headquarters, Washington, DC; and the members of the STS-61 flight crew, commander Richard O. Covey, pilot Kenneth D. Bowersox, and mission specialists Tom Akers, Jeffrey A. Hoffman, F. Story Musgrave, Claude Nicollier (European Space Agency), and Kathryn C. Thornton.

14. Ron Sheffield, email to author, June 5, 2018.

15. The repair of the Space Telescope Imaging Spectrograph on the fourth servicing mission is a good example. To get the lid off the instrument, astronauts Mike Massimino and Mike Good had to remove over one hundred noncaptive screws and washers. Engineers at the Goddard Space Flight Center developed a device called the Fastener Capture Plate that ensured these would not float away in the process (https://www.spacetelescope.org/images/hst3-venice2010/) and a new high-speed, low-torque power tool to speed up the task without the risk of stripping a screw. A special card extraction tool was developed to overcome the challenge of grabbing a single circuit board with spacesuit gloves, which is roughly equivalent to picking a single sheet of paper out of a stack while wearing ski gloves. See also https://www.spacetelescope.org/about/history/stis_repair/ and https://www.spacetelescope.org/about/history/tools/.

BIBLIOGRAPHY

Bahcall, John N., and Lyman Spitzer, Jr. "The Space Telescope." *Scientific American* 247, no. 1 (July 1982): 40–51.

Bilstein, Roger E. *Stages to Saturn: A Technological History of the Apollo/Saturn Launch Vehicles*. NASA SP-4206. Washington, DC: National Aeronautics and Space Administration, 1996.

"A Brief History of the Hubble Space Telescope." *Quest: The History of Spaceflight Quarterly* 17, no. 2 (2010): 6–17.

Brown, Nelson E. *Application of EVA Guidelines and Design Criteria*. Vol. 1, *EVA Selection/Systems Design Considerations*. Vol. 2, *EVA Workstation Conceptual Designs*. Houston, TX: NASA Lyndon B. Johnson Space Center, 1973.

Camp, David W., D. M. Germany, and Leonard S. Nicholson. *STS-31 Space Shuttle Mission Report*. NSTS-08207. Washington, DC: National Aeronautics and Space Administration, 1990.

Chaisson, Eric. *The Hubble Wars; Astrophysics Meets Astropolitics in the Two-Billion-Dollar Struggle Over the Hubble Space Telescope.* New York: Harper Collins, 1994.

Devorkin, David H. *Science with a Vengeance: How the Military Created the U.S. Space Sciences after World War II*. New York: Springer, 1992.

Feynman, Richard P. "Mr. Feynman Goes to Washington." *Engineering and Science* 51, no. 1 (1987): 6–22.

Field, George B., and Donald Goldsmith. *The Space Telescope: Eyes above the Atmosphere*. Chicago: Contemporary Books, 1989.

Fisher, Arthur. "The Trouble with Hubble." *Popular Science* 237, no. 4 (October 1990): 72–76, 100.

Fisher, William F., and Charles R. Price. *Space Station Freedom External Maintenance Task Team—Final Report*, vol. 1, parts 1–2. Houston, TX: NASA Lyndon B. Johnson Space Center, 1990.

Gainor, Chris. *Hubble Space Telescope Operating History*. Washington, DC: NASA History Series, forthcoming.

Gieryn, Thomas F., and Anne E. Figert. "Ingredients for a Theory of Science in Society: O-Rings, Ice, Water, C-Clamp, Richard Feynman and the New York Times." In *Theories of Science in Society*, ed. Susan Cozzens and Thomas F. Gieryn, 67–97. Bloomington: Indiana University Press, 1991.

Gillette, Estella Hernandez. "A Case Study of the Socialization Processes of the NASA Spacewalkers in the High Reliability Organizational Culture of the Extravehicular Activity (EVA) Teams." PhD diss., George Washington University, 2013.

Graham, Stephen, and Nigel Thrift. "Out of Order: Understanding Repair and Maintenance." *Theory, Culture and Society* 24, no. 3 (2007): 1–25.

Green, Constance McLaughlin, and Milton Lomask. *Vanguard: A History*. NASA SP-4202. Washington, DC: National Aeronautics and Space Administration, 1970.

Grumman Aerospace Corporation. "OAO/LST Economics Study." NASA contract NAS5-17149, October 1, 1970. Space Telescope History Project, Acc. 1999-0035, National Air and Space Museum, Smithsonian Institution, Box 31.

Grumman Aircraft Engineering Corporation. "The OAO in Association with a Manned Space Station." NASA contract NASW-1143, 1965. Space Telescope History Project, Acc. 1999-0035, National Air and Space Museum, Smithsonian Institution, Box 31.

Hale, Wayne N. "Flying the Shuttle: Operations from Preparation through Flight to Recovery." In *Space Shuttle Legacy: How We Did It and What We Learned*, ed. Roger D. Launius, John Krige, and James I. Craig, 173–189. Reston, VA: American Institute of Aeronautics and Astronautics, 2013.

Harwood, William. "How NASA Fixed Hubble's Flawed Mirror—and Reputation." *CBS Interactive*, April 22, 2015. https://www.cbsnews.com/news/an-ingenius-fix-for-hubbles-famously-flawed-vision/.

Hersch, Matthew. *Inventing the American Astronaut*. New York: Palgrave Macmillan, 2012.

Kinge-Hele, D. G., and Eileen Quinn. "The Variation of Upper-Atmosphere Density between Sunspot Maximum (1957–1958) and Minimum (1964)." *Journal of Atmospheric and Terrestrial Physics* 27, no. 2 (February 1965): 197–209.

Launius, Roger D., and David DeVorkin. *Hubble's Legacy: Reflections of Those Who Dreamed It, Built It, and Observed the Universe with It*. Washington, DC: Smithsonian Institution Scholarly Press, 2015.

Launius, Roger D., John Krige, and James I. Craig, eds. *Space Shuttle Legacy: How We Did It and What We Learned*. Reston, VA: American Institute of Aeronautics and Astronautics, 2013.

Lockheed Missiles and Space Corporation. "Final Design Review Briefing, Support Systems Module Phase B. DR-MA-03, DPD 449." February 11, 1976. Private collection of Ronald L. Sheffield.

Logsdon, John. *After Apollo? Richard Nixon and the American Space Program*. New York: Palgrave Macmillan, 2015.

Longair, Malcolm S. "The Space Telescope and Its Opportunities." *Quarterly Journal of the Royal Astronomical Society* 20 (February 1979): 5–28.

Marshall Space Flight Center. "ST Support Systems Module Request for Proposals. NAS8-1-6-PP-00596Q." 1977. Space Telescope History Project, Acc. 1999-0035, National Air and Space Museum, Smithsonian Institution, Box 32.

Massimino, Michael J. *Spaceman: An Astronaut's Unlikely Journey to Unlock the Secrets of the Universe*. New York: Crown Archetype, 2016.

Mattice, James J. *Hubble Space Telescope Systems Engineering Case Study*. Dayton, OH: Center for Systems Engineering, Air Force Institute of Technology, Wright-Patterson AFB, 2005.

McDonnell Douglas Technical Services Company. "Application of Shuttle EVA Systems to Payloads," Report MDC W00014," June 1976.

Mindell, David A. *Our Robots, Ourselves*. New York: Viking Press, 2015.

Murphy, J. T., and C. R. Darwin. "Lower Payload Costs through Refurbishment and Module Replacement." *Astronautics and Aeronautics* 11 (May 1973): 40–47.

National Aeronautics and Space Administration. *The Space Telescope*. SP-392. Washington, DC: National Aeronautics and Space Administration, 1976.

National Research Council. *Space Research: Directions for the Future*. Washington, DC: The National Academies Press, 1966.

Neal, Valerie. *Designing a Telescope for Servicing in Space*. Huntsville, AL: Marshall Space Flight Center, 1984.

Neal, Valerie. *Spaceflight in the Shuttle Era and Beyond: Redefining Humanity's Purpose in Space*. New Haven: Yale University Press, 2017.

Neufeld, Michael J., and John B. Charles. "The Invention and Diffusion of Neutral Buoyancy Training." In *History of Rocketry and Astronautics: Proceedings of the Forty-Ninth History Symposium of the International Academy of Astronautics*, ed. Tal Inbar, 59–71. San Diego: Univelt, Inc., for the American Astronautical Society, 2017.

Newman, Ronald L. *STS-61 Mission Director's Post-Mission Report*. NASA Technical Memorandum 104803. Houston, TX: NASA Lyndon B. Johnson Space Center, 1995.

Overbye, Dennis. 1993. "Hubble Jeopardy." *New York Times*, November 28.

Rogers, William P., Neil A. Armstrong, David C. Acheson, Eugene E. Covert, Richard P. Feynman, Robert B. Hotz, Donald J. Kutyna, Sally K. Ride, Robert W. Rummel, Joseph F. Sutter, Arthur B. C. Walker, Jr., Albert D. Wheelon, and Charles E. Yeager. *Report of the Presidential Commission on the Space Shuttle Challenger Accident*. Washington, DC: National Aeronautics and Space Administration, 1986.

Ross-Nazzal, Jennifer. "An Interview with Kathy Sullivan." *Quest: The History of Spaceflight Quarterly* 17, no. 2 (2010): 18–33.

Rumerman, Judy A. *NASA Launch Systems, Space Transportation, Human Spaceflight and Space Science 1979–1988*, vol. 5: *NASA Historical Data Book*, SP-4012. Washington, DC: National Aeronautics and Space Administration, 1999.

Scott, David Meerman, and Richard Jurek. *Marketing the Moon: The Selling of the Apollo Lunar Program*. Cambridge, MA: MIT Press, 2014.

Shapin, Steven. "The Invisible Technician." *American Scientist* 77, no. 6 (1989): 554–563.

Shayler, David J., and David M. Harland. *The HST: From Concept to Success*. New York: Springer Praxis, 2016.

Sherr, Lynn. *Sally Ride, America's First Woman in Space*. New York: Simon and Shuster, 2014.

Smith, David A. *Designing an Observatory for Maintenance in Orbit: The Hubble Space Telescope Experience*. Huntsville AL: Space Telescope Project Office, NASA Marshall Space Flight Center, 1986.

Smith, Robert W., with Paul A. Hanle, Robert H. Kargon, and Joseph N. Tatarewicz. *The Space Telescope: A Study of NASA, Science, Technology and Politics*. Cambridge: Cambridge University Press, 1989.

Spitzer, Lyman, Jr. *The Astronomical Advantages of an Extra-terrestrial Observatory*. Project RAND Report, Douglas Aircraft Company, September 1, 1946.

Spitzer, Lyman, Jr. "History of the Space Telescope." *Quarterly Journal of the Royal Astronomical Society* 20 (1979): 29–36.

Styczynski, Thomas E. "Hubble Space Telescope First Servicing Mission." *SAE Transactions* 103 (1994): 2072–2076.

Tatarewicz, Joseph N. "The Hubble Space Telescope Servicing Mission." In *From Engineering Science to Big Science: The NACA and NASA Collier Trophy Research Project Winners*, ed. Pamela E. Mack, NASA SP-4219, 365–396. Washington, DC: National Aeronautics and Space Administration, 1998.

Wilde, Richard C., James W. McBarron II, Scott A. Manatt, Harold J. McMann, and Richard K. Fullerton. "One Hundred US EVAs: A Perspective on Space Walks." *Acta Astronautica* 51, nos. 1–9 (2002): 579–590.

Zimmerman, Robert. *The Universe in a Mirror: The Saga of the HST and the Visionaries Who Built It*. Princeton, NJ: Princeton University Press, 2008.

ORAL HISTORY INTERVIEWS

Adams, Trevino, Havens. Interview by Joseph N. Tatarewicz, January 9, 1986. Space Telescope History Project, Acc. 1999-0035, National Air and Space Museum, Smithsonian Institution, Box 15, Folder 1.

Cepollina, Frank J. Interview by Sandra Johnson, June 11, 2013. NASA Headquarters Oral History Project. https://www.jsc.nasa.gov/history/oral_histories/NASA_HQ/Administrators/CepollinaFJ/CepollinaFJ_6-11-13.htm.

Costa, Frank V. Interview by Joseph N. Tatarewicz, March 31, 1989. Space Telescope History Project, Acc. 1999-0035, National Air and Space Museum, Smithsonian Institution, Box 16, Folder 7.

Costa, Frank V. Discussion with the author, October 2017.

Downey, James A. Interview by Robert W. Smith, January 18, 1984. Space Telescope History Project, Acc. 1999-0035, National Air and Space Museum, Smithsonian Institution, Box 16, Folder 6.

Leckrone, David A. Interview by Robert W. Smith, August 14, 1984. Space Telescope History Project, Acc. 1999-0035, National Air and Space Museum, Smithsonian Institution, Box 18, Folder 3.

McCandless, Bruce C. II. Interview by Joseph N. Tatarewicz, January 8–9, 1986. Space Telescope History Project, Acc. 1999-0035, National Air and Space Museum, Smithsonian Institution, Box 18, Folder 17.

Nein, Max E. Interview by Robert W. Smith, June 5, 1984. Space Telescope History Project, Acc. 1999-0035, National Air and Space Museum, Smithsonian Institution, Box 18, Folder 11.

Odom, James. Interview by Andrew Dunar and Stephen P. Waring, August 26, 1993. NASA Marshall Space Flight Center History Archives. https://www.nasa.gov/sites/default/files/atoms/files/19930826_james_odom_.pdf.

Olivier, Jean R. Interview by Stephen P. Waring, July 20, 1994. NASA Marshall Space Flight Center History Archives. https://www.nasa.gov/sites/default/files/atoms/files/19940720_jean_olivier_oral_history_interview.pdf.

Wojtalik, Fred S. Interview by Stephen P. Waring, July 20, 1994. NASA Marshall Space Flight Center History Archives. https://www.nasa.gov/sites/default/files/atoms/files/19940720_fred_wojtalik_oral_history_interview.pdf.

INDEX

Note: Page numbers in italic type indicate illustrations.